With 60% of the land in the United States under private ownership, the role of the private sector in the conservation of habitat and species diversity is being recognized as increasingly important. *Dealing in diversity* examines the 'market' for conservation of natural areas in the USA, considering the efforts of both profit and nonprofit making ventures and discusses the costs and benefits of protecting natural areas, using specific examples of landowners and agencies involved in private sector conservation. It concludes by discussing the potential and limitations of the private conservation market and the role of the government in that market. The effectiveness of conservation methods is examined at three levels: constitutional, organizational and operational. This book will therefore appeal to all those interested or involved in conservation, from students to policymakers.

DEALING IN DIVERSITY: AMERICA'S MARKET FOR NATURE CONSERVATION

DEALING IN DIVERSITY: AMERICA'S MARKET FOR NATURE CONSERVATION

VICTORIA M. EDWARDS

*Senior Lecturer in Land Management, Department of Land and
Construction Management, University of Portsmouth*

CAMBRIDGE
UNIVERSITY PRESS

CAMBRIDGE UNIVERSITY PRESS
Cambridge, New York, Melbourne, Madrid, Cape Town, Singapore, São Paulo, Delhi

Cambridge University Press
The Edinburgh Building, Cambridge CB2 8RU, UK

Published in the United States of America by Cambridge University Press, New York

www.cambridge.org
Information on this title: www.cambridge.org/9780521117579

First published 1995
This digitally printed version 2009

A catalogue record for this publication is available from the British Library

Library of Congress Cataloguing in Publication data
Edwards, Victoria M.
Dealing in diversity : America's market for nature conservation /
Victoria M. Edwards.
p. cm.
Includes bibliographical references (p.   ) and index.
ISBN 0 521 46567 2 (hc)
1. Nature conservation – United States.   2. Natural areas – United
States.   3. Nature conservation – Economic aspects – United States.
4. Natural areas – Economic aspects – United States.   I. Title.
QH76.E34   1995
333.78′216′0973 – dc20   94–36811 CIP

ISBN 978-0-521-46567-0 hardback
ISBN 978-0-521-11757-9 paperback

To George and Betty Edwards
for their love and forbearance

Contents

Preface

This book is about nature conservation – a topic that has increasingly captured the attention of academics, practitioners and the general public during the 1990s. However, while there has been much debate about the need to conserve natural areas and biological diversity and about the technical means of carrying out such conservation, attention has only just begun to turn to the economic, legal and institutional problems of conservation. It seems that the 64-million-dollar question is, how can we make nature conservation pay?

Innovative means of paying for nature conservation are emerging in the USA through the development of a private sector conservation 'market'. *Dealing in diversity* examines the efforts of both profit and nonprofit making ventures and evaluates their success in combining cost effectiveness with sound conservation practice. The book comprises a carefully constructed mixture of theory and practice: examining economic and political theories of conservation, providing a diverse collection of interesting and instructive case studies and a valuable introduction to the operation of the conservation market by explaining the relevance of economic theory, organizations and legal and financial mechanisms for protecting natural areas. It evaluates the potential for maximizing conservation benefits while minimizing environmental costs by examining case studies, including fee-hunting, watchable wildlife, conservation real estate and landowner associations. Throughout its analysis of the conservation market the following points are specifically addressed:

the definition and development of conservation products;
clarification of the roles of the parties involved; and
improvement of the institutional arrangement of the market.

The book concludes by answering three questions: (a) how much can the conservation market contribute to the protection of America's natural areas?;

(b) can improved efficiency of the market increase its capacity to deal with present and future problems?; and (c) what, if any, is the role for government in the private conservation market?

Dealing in diversity is intended for use by students of the social and environmental sciences and by policymakers, conservationists and land managers. It draws on experience of American, British and New Zealand conservation provision to explain the private sector conservation market in the USA. A multidisciplinary approach to conservation is adopted, combining basic biological understanding with economics, political theory, law, valuation and management. It is intended to be accessible to those with no specialist training in any of these disciplines.

The first four chapters of the book provide a detailed examination of the operation of the conservation market. Chapter 1 introduces the concept of private sector conservation of natural areas and outlines the advantages of a private market for conservation. The chapter suggests that if there is demand for more natural areas, then innovative means of providing them through the private sector should emerge. Chapter 2 identifies the costs and benefits (and specifically their distribution) of protecting natural areas. It examines the characteristics of the conservation market and analyzes the type of benefits derived from natural areas. The chapter explains that the institutional arrangement of the market establishes rules and provides incentives that influence the decisions of individuals. The government can assist the market by providing formal mechanisms for protecting natural areas, but can also hinder development of the market by allowing the persistence of inflexible and inappropriate institutional arrangements. Throughout the book examples of such hindrances are highlighted. Chapter 3 identifies private conservation organizations that have been established to act as agents between landowners and the public, focusing on The Nature Conservancy (the USA's largest organization that directly protects private land) and Land Trusts (small, local trusts, of which USA has around 900). Chapter 4 examines the formal protection mechanisms used to ensure continued conservation of natural areas through purchase, bargain sale or donation of certain property rights. It evaluates the relative advantages and disadvantages of each mechanism and identifies the incentives available to landowners entering into conservation agreements.

Chapters 5 to 8 focus on a variety of profit and nonprofit-making ventures in the private conservation sector. Chapter 5 provides an examination of the demand for conservation based recreational activities in the USA and includes specific analysis of the demand for fee-hunting on private land. It provides detail of the management techniques used to supply fee-hunting and evaluates the potential costs and benefits of fee-hunting enterprises. Chapter

6 examines nonconsumptive recreational use of wildlife. It identifies how demand might be met with private sector supply and might subsequently allow for the collection of revenue for conservation areas. It concludes by explaining how public access to wildlife must be carefully managed in order to minimize environmental costs and emphasizes that private management coupled with restricted access is, therefore, compatible with protection. Chapter 7 examines the specialist 'conservation real estate' market which has arisen in the USA, whereby a legally protected area is marketed as a 'conservation property'. It illustrates how real estate agents and conservation organizations can combine to sell a mutually beneficial product and identifies the concept of 'partial development', whereby a conservation organization protects part of a property and allows development of noncritical areas. It examines the management techniques used to enable partial development and conservation to coexist and concludes with an examination of the tax system and its potential for encouraging development rather than conservation. Chapter 8 acknowledges the willingness that exists amongst some landowners to bear the costs of restoring and protecting natural areas. It examines how a group of neighboring landowners might combine through some formal voluntary agreement, to self-regulate the management of a natural area, and explains how privately created and enforced regulation can facilitate the protection of natural areas of land in multiple ownership. The chapter concludes by illustrating how public sector interference in these private sector arrangements can destroy the benefits created.

The concluding chapter of the book explains how organizations involved in the protection of natural areas are seeking to widen their brief and adopt a more holistic approach to conservation, integrating the management of natural areas with the economic use of neighboring land. It warns that in attempting to broaden their own brief, conservation organizations are in danger of losing the very essence of their success – their private approach. The chapter concludes by questioning the respective roles of the private and public sectors in nature conservation.

Dealing in diversity was never intended to be a book that justifies a free market approach to public goods provision by citing successful *private* conservation examples. Instead, its purpose is to correct the balance of the plethora of conservation books that only identify the role of *public* sector in conservation provision. The book serves as a testimony to the valuable conservation work that *can* be achieved by a private sector approach. While the limitations of this approach are explained, the case studies included in this book provide more a directory of best practice. Although research has not uncovered any specific examples of 'bad' conservation enterprises (where financial exploitation of the natural area is carried out to the detriment of conservation), such examples inevitably exist. Just as the government can falter in its provision of nature conservation (and a few examples are cited in the book), so too can the private sector. However, we must be careful not to label poor private sector examples as 'failures', a term which suggests irreparable damage. Instead, it is better that we continually seek to improve private sector provision and its effectiveness in providing an economic alternative to public sector provision.

Victoria M. Edwards

Acknowledgments

This project would not have been possible without fellowships from the Winston Churchill Memorial Trust and the Political Economy Research Center in Montana. My sincere thanks to both organizations for their financial support and encouragement.

Eventual responsibility for the theory and practical applications of nature conservation and the way in which they are communicated in this book must rest, of course, with the author. However, a huge number of people have contributed to this work by accommodating me and my lines of enquiry. As if putting up with my visits weren't enough, most of them also offered extensive help in reviewing the work and updating information. They hold the common characteristic of striving to improve the management of America's natural areas. I hope that I have done their cause justice. In particular, my thanks to: Terry Anderson, Kathy Barton, Chris Boyd, Nancy Braker, John Brantley, David Campbell, David Carr, Lane Coulston, Rick Danver, Ron Geatz, Sandy Goodman, Anne Gore, Brent Haglund, Chuck and Katie Harrison, Marty Jessen, Don Leal, Bill Long, Wayne Long, Russell Miller, Will Murray, John Nelson, Martha Noble, Deborah Osborne, Dave Richie, Deborah Richie, Mary Sexton, Greg Simonds, Fred Smith, John Stokes, Robert Streeter, Tony Thompson, Steven Thorne, Barry Truit, Ralph Waldt, Cathy Walters, Grant Werschkull, Zach Willey, Brian Winter. My special thanks to Lisa Mueller and Ron Stanley who helped to arrange a stimulating and enjoyable program of meetings for me. The late Jim Thompson of North Heron Lake continuously proved a source of inspiration and encouragement, for which I am very grateful. Jim was very keen for this work to be published and I am truly sorry that he didn't get to see it in print.

I am indebted to Bill Seabrooke for reading and criticizing parts of the manuscript and to colleagues at Portsmouth for their support in completing the book. A special thanks to Helen Pickering, who went well beyond the call of duty in offering extensive and helpful comments on every chapter and to Ruth Pearson and Chris Evans who meticulously prepared and reprepared the illustrations. Several individuals have generously provided photographic illustrations. These contributions are detailed in the photographic captions. My thanks to Alan Crowden and Myra Givans for guiding me through the publication process. Responsibility for all errors, omissions and shortcomings is, of course, my own.

1

Introduction: private provision of conservation

> There is a clear tendency in American conservation to relegate
> to government all necessary jobs that private landowners fail to
> perform.
>
> *Aldo Leopold (The Land Ethic)*

Background

When the North American continent was first settled in the sixteenth and
seventeenth centuries, its land, undeveloped and rich in natural resources,
supported a sparse population of some 12 million American Indians. The
first colonists from England, who had left behind land scarcity and rising
land prices, found a great contrast on the east coast of America:

the virgin American forests and their wildlife populations including white tailed deer,
wild turkey, bobcats, cougars, ruffled grouse, black bear and wolves extended almost
continuously from the East coast to the prairies of the Midwest. Salmon and shad
migrated up the major Atlantic coastal rivers to spawn. The grassland prairies and
plains stretching to the foothills of the Rocky Mountains were populated with unnum-
bered bison and prong-horned antelope, prairie birds, elk, mule deer, wolves, and
grizzly bears. The lands beyond the Rockies to the Pacific coast were occupied by
the crests of bare mountains at the higher elevations and deserts at the lowest. The
lands between the deserts and the mountain tops were covered with evergeen forests
and relatively dry grasslands. The great valley of central California was a vast plain
and marshlands harboring elk, prong-horns, salmon, and grizzly bears. America was
a land of natural beauty as well as abundant natural resources.

Council on Environmental Quality, 1984

Currently, the USA supports almost 270 million Americans. Since it was
first settled, more than 90% of the tallgrass prairies, 55% of wetlands, 26%
of all forests, 50% of tropical forests and 75% of the old growth forests have
been destroyed (The Nature Conservancy, 1990; World Resources Institute,
1990). This loss of habitat is reflected in subsequent reduction in the number
of species of native flora and fauna. Since 1620, over 500 species and sub-
species of native plants and animals have become extinct in North America
(Thatcher, 1988). Casualties include the passenger pigeon (*Ectopistes
migratorius*), Labrador duck (*Camptorhynchus labradorius*), Carolina

1

parakeet (*Conuropsis carolinesis*), heath hen (*Tympancuchus cupido cupido*) and the plains bison (*Bison bison*).

The extinction of species and subsequent decline in biological diversity has begun to cause worldwide concern. Some scientists estimate that around 1000 species become extinct each year and conjecture that this rate of extinction could rise to 10 000 species per annum during the 1990s (Thatcher, 1988). According to estimates by Raven (1988), the earth may be losing 100 species per day, while Wilson (1988) and Myers (1986) estimated a loss of 50 species per day. Although the extent of loss is disputed, it is generally recognized that species extinction is proceeding at an increasing rate, which is inconsistent with evolutionary trends. Alarmed at the accelerated loss, nations are searching for means to halt this dramatic reduction in biological diversity. In 1983, the United Nations established an independent body to reexamine the critical environment and development problems of the world. In 1987, The World Commission on Environment and Development, chaired by Gro Harlem Brundtland (then Prime Minister of Norway), reported that: 'A first priority is to establish the problem of disappearing species and threatened ecosystems on political agendas as a major economic and resource issue.' (World Commission on Environment and Development, 1987: 13).

The causes for the loss of species are numerous, including, for example, over-exploitation of plant and animal species, the impact of exotic species, the degradation of habitat through pollution and may soon include climatic change. However, it is widely recognized that the most important cause is the loss and fragmentation of natural habitats. The recent willingness of governments to address these environmental issues and take on the recommendations of the Brundtland report reflects not only a greater appreciation on their part of the rate of destruction of habitat but also their perception of an enlightened public interest in the environment.

Such concern for the protection of natural areas has been present in the USA throughout the majority of this century. Early efforts concentrated on protecting habitats supporting threatened species, with conservationists often using the metaphor of an ark as justification. Protecting such species and biological diversity ensures, like Noah, that nothing is lost from the full complement of natural resources that have evolved on earth. Naturalist Aldo Leopold identified that 'the first prerequisite of intelligent tinkering is to save all of the cogs and wheels.'

The environmental debate has since developed, with scientists, economists, conservationists and civil servants all contributing and designing policies aimed at reflecting the enlightened public interest in land resources and in the management of the nation's lands. In the midst of this debate, landowners are finding themselves the holders of the key resource – the property rights to land. Whereas in the past habitat conservation in the USA focused on the retention and management of *public* land, both federal and state, more recently it has started to address the role of the *private* landowner in the protection of natural areas.

The importance of private land

There are three principal reasons why private land must be included in analysis of the conservation movement in the USA. First, the very extent of the private domain suggests that the protection of wildlife habitat on private land is important. Approximately 40% of the land area of the USA is in government ownership (federal, state and local), with the majority of federal land located in Alaska and the western states. Only 4% of the area of the Great Plains and the eastern states lies in federal ownership. The geographic disparity is also manifest at the state level. Within each state, the ratio of public land to private land varies considerably: public ownership ranges from 0.3% of land in Connecticut to 86.5% of land in Nevada (Bandow, 1986). The extent of public ownership leaves a total of 60% of the area of the USA in private ownership, with even greater concentrations of private estates in the east.

Second, it is important to recognize the type of land which is held publicly: 'Much of the public land, while incomparable in scenic and natural splendor, is relatively sparse as a pool of biological and genetic diversity.' When the west was settled, the biologically rich and important lands (often around wetland areas, streams and rivers) were the most popular sites for homesteading by private individuals: 'The vast extent of the Bureau of Land Management's holdings in the West are generally arid, dominated by sagebrush and a few other species, and are relatively impoverished in wildlife. Many of the extensive areas of national forest in the West are also characterized by relatively few plant and animal species.' (Council on Environmental Quality, 1984: 364). Many rare and valuable types of habitat may be found only on private land. For example, the federal agency most commonly associated with the conservation of wetland areas (the Fish and Wildlife Service) controls less than 5% of the nation's wetlands. Systematic inventories conducted by the natural resource agencies of various states suggest that a variety of natural habitats has survived on private land, supporting a number of rare and endangered species. For example, the US Department of the Interior and the US Fish and Wildlife Service (USDI and USFWS, 1990) report that 74% of the USA's wetlands are in private ownership: in Louisiana more than 85% of the wetland habitat is privately owned (Louisiana Private Lands Team, 1989).

Third, much of the wealth of wildlife species on public lands is located in national parks or national wildlife refuges. These areas tend to be concentrated in mountainous regions, where natural physical restrictions limit land-use conflicts. To afford proper protection, these parks and reserves depend upon sound management practices not only within their confines, but also on neighboring land. The development pressures often found on lands surrounding the parks can create an 'island' refuge situation, where the area is too small to sustain wildlife over the longer term. In addition, protected habitats can be starved of essential resource inputs, such as water, by the management practices adopted on adjacent lands. Water diversion on land outside the Everglades National Park has severely disrupted the Park's ecosystem (Whelan, 1991). The viability of small protected areas, in particular, can be jeopardized by the breakdown of the larger systems on which they depend. Research by Newmark (1987) has suggested that, all variables being equal, large areas provide more secure protection for species than do small areas (see Fig. 1.1). Reflecting this, the concept of protecting biological diversity through protecting the wider ecosystem, rather than specific species, has been the topic of much recent research. Early work by Craighead (1979) identified that the needs of the grizzly bear (*Ursus arctos horribilis*) in Yellowstone National Park could not be met within the confined boundaries of the park. Recent work on the Yellowstone Greater Ecosystem concept advocates that 68 million acres of land be integrated into some sort of management strategy in order to provide the primary habitat necessary to sustain all native species in the region. Yellowstone National Park currently comprises some 2.2 million acres. At the World Congress on National Parks, consensus was reached on the need for the protected areas in the world to be at least tripled to fairly represent a sample of the Earth's ecosystems (McNeely & Miller, 1984).

While there has been widespread recognition that attention must be focused on private land to achieve adequate wildlife protection, there remains much debate as to the best means of achieving protection on such land. The debate often focuses on two distinct alternatives: (a) public sector incorporation or regulation and (b) private sector provision.

Public sector provision

Public ownership of land

Public sector provision of protected habitat can be achieved through two types of policy. First, governments can acquire the property rights to the land, adding the land to the public domain and thereafter managing it as a public

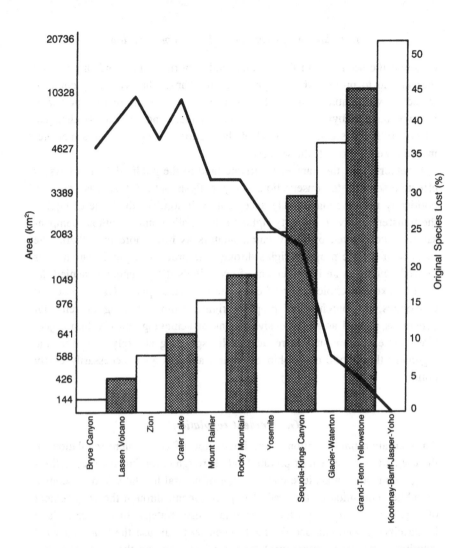

Fig. 1.1. Habitat area and loss of large animal species in North American National Parks 1986. Source: Newmark (1987).

resource. This is seen as the ultimate form of conservation control by some policymakers, with freehold ownership ensuring that the desired form of management can be fully implemented by the appropriate public agency. Without the need to negotiate terms of management with a private landowner, the agency can decide upon the appropriate use(s) of the natural habitat and plan its future management to achieve stated conservation objectives. Under this option, effective habitat protection is dependent on the appropriateness of the agency's mandate and the interests of its stakeholders. Foresta (1984) argued that the US National Park Service, with its multipurpose mandate 'to conserve the scenery and the national and historic objects and the wildlife therein and to provide the enjoyment of the same,' has variously compromised between habitat conservation and recreation, as the relative power of its two main lobbying groups (environmentalists and recreators) changed. Chase (1987) was equally critical of the National Parks Service's management of Yellowstone National Park.

Whether or not the transfer of private land to the public domain offers an effective solution to conservation, many analysts believe that it is neither a politically nor an economically acceptable solution. On the issue of equity, the transfer of private property rights to the public domain without compensation is considered by many commentators as little more than theft. The transfer of private property rights through the process of public purchase is viewed equally as an infringement of civil rights if the vendor is denied the right to exercise choice over whether the sale takes place. On the issue of economics, the purchase of property rights, from a willing or reluctant vendor, is viewed as an expensive means of achieving conservation objectives. Increased pressure to reduce public spending is likely to further the argument that public ownership of natural areas is an unnecessarily costly option.

Government regulation

An alternative policy option is the imposition of government regulation on the use and management of private land. Through enforcing such regulations the government can achieve conservation of natural habitat without recourse to public acquisition of the land. The government, through the enforcement of regulations, can control the specific land-use strategy for a piece of land by specifying who can use the land, when they can use the land and what activities (farming or otherwise) they can carry out on the land. The proponents of government regulation argue that centralized control of land use is the only means of achieving conservation objectives: 'environmental problems cannot be solved through cooperation . . . and the rationale of government with major coercive powers is overwhelming' (Ophuls, 1973). Other commentators raise several issues to advance criticism of public sector provision. For example, Burton (1978) stated that in advocating government intervention there has been an implicit assumption that the cost of administering that intervention is zero. Clearly, this is unlikely to be true in the case of the protection of private land. A public agency must assess the amount of land and the specific areas of land to be protected, develop a formal protection mechanism and negotiate, monitor and enforce that agreement with the private landowner. Indeed, the costs of government intervention may outweigh the benefits. A further criticism of the reliance on government provision centers around the unrealistic assumption that both government electees and public servants will aim to maximize social efficiency in their design and implementation of public policy for conservation. In the case of political appointees, we might assume that decisions will be determined by vote considerations, with powerful lobbying (which may not accurately represent public opinion) likely to sway decision-making. In addition, even if the majority rule of voting is accurately used, social welfare will not necessarily be maximized (Buchanan, 1973). It can also be said that public servants responsible for the management of conservation projects will have their own goals. Baden & Stroup (1983) criticized the US Forest Service for selecting areas for timber extraction on the basis of political consideration rather than site productivity.

Threats of increased regulation over the use of land and punitive measures for noncompliance with the government's conservation objectives have begun to cause unease in the agricultural industry. Landowners and farmers have expressed concern over the constraints that conservation policy initiatives may impose on the management of their land. This may well lead to a negative attitude towards the issue of conservation and a reluctance to contribute to the design of conservation policy. Conservation is already considered by landowners as the external imposition of significant private cost. If regulation is the way ahead, then there will have to be changes made to this perception.

Coase (1960) argued that many kinds of environmental problems can be taken care of under a suitable system of property rights. He suggested that contractual agreements (accompanied by payment) between a private producer and other individuals who may gain or lose from the producer's activity, could result in socially optimal behavior on the part of the producer. The 'Coase Theorem' resulting from the work implies either that the private producer may compensate the individuals who may suffer from his action or that the individuals may pay the producer to take their interests into account.

If the Coase Theorem holds true, then in order to increase the amount of land in private ownership protected as wildlife habitat, society should enter into transactions that increase the benefits available to landowners and/or reduce the costs they incur through protection.

Private sector solutions

A growing number of resource economists are proposing market solutions to contemporaneous environmental problems (see, for example, Anderson &

Leal, 1991). The retail industry is a good example of an industry intent on gaining from the general increase in environmental awareness. By uniting their own objectives of increased sales with the consumer's demand for green products, the retailers have successfully harnessed increased public awareness in the environment for their own benefit. Supermarkets have stocked their shelves with ozone-friendly aerosols, dolphin-friendly tuna and river-friendly detergents. Changes are not only product based. In an attempt to improve the operation of the market, increased information is available on the contents of products for sale and their effect on the environment, and an advocacy role has also been adopted, with shops encouraging the consumer to reuse carrier bags – good economic sense for a supermarket expected to provide free bags.

While other industries, such as publishing, retailing and, more recently, tourism have created the concept of green consumerism, so too are landowners beginning to develop a market for conservation. Wise landowners are adopting a new, positive and private approach to conservation and are adding value to their estate by harnessing the benefits to be derived from protection of natural areas. The private sector in conservation is made up of a number of these individuals and groups, each with their own demands and approaches to protecting natural areas. In the establishment and management of conserved areas, it is possible for landowners to combine their own private objectives with the conservation objectives of the general public. In doing so, conserved natural areas could be developed successfully as 'products'.

Until recently, the 'market' for nature conservation has been very limited: operating through the property market, it has offered opportunities only to landowners and farmers who wish to dispose of their assets. The market has done little to bring together the holders of property rights with those who wish to purchase conservation without acquiring the freehold rights to land, or indeed the costs which accompany them. However, the conservation market is developing and a more sophisticated private market for conservation is evolving in America, responding both to the demand for natural areas and biological diversity and also to the specific nature of the supply: the land and restrictions imposed by land use. The task of the landowner now is to 'sell' the conserved area without necessarily selling the freehold property rights to the land. Developing conserved areas into marketable products involves understanding the specific nature of the site and the benefits to be derived from it, and finding mechanisms by which those benefits might be transferred to willing purchasers. Some benefits can be derived from the use of the area, other benefits may be derived from the development and sale of associated, tangible products. Purchasers of such products could include private individuals and groups, corporations or public bodies.

Proponents for the market approach argue that well-defined, enforceable and transferable property rights, accompanied by economic incentives, will enable entrepreneurs to meet the conservation demands of society while furthering their own self-interest. They believe that government intervention in the market place should be seen as a last resort, only to be pursued when definition and enforcement of property rights are either impossible or extremely costly. Even at this stage, government intervention is viewed as unfortunate: 'those kinds of solutions often become entrenched and stand in the way of innovative market processes that promote fiscal responsibility, efficient resource use, and individual freedom.' (Anderson & Leal, 1991).

Identifying a role for government

The Organisation for Economic Cooperation and Development, in its study *Renewable Natural Resources: Economic Incentives for Improved Management* (OECD, 1989), characterized the causes for 'policy failures' in a variety of countries as ' "government failures" attributable to public ownership and management of natural resources and the broad range of government incentives aimed at private owners, or "market failures" derived from private ownership and exploitation.' The dichotomy expressed typifies an approach to natural resource management issues that ignores a broad range of possible solutions not characterized as a purely public or purely private approach. Just as some analysts assume that government provision offers a panacea to the problem of providing for nature conservation, so others offering market solutions often ignore the problems encountered in establishing and maintaining a well-functioning market.

Some authors recognize the need to evaluate not only the goods and services that are produced by the market, but also the market itself and the institutional arrangements that govern its operation. In particular, analysis must reveal the extent to which there is a role for government in assisting in the development and enhancement of the market. Ostrom (1990) argued that 'no market can exist for long without underlying public institutions to support it. In field settings, public and private institutions frequently are intermeshed and depend on one another, rather than existing in isolated worlds.'

Taking this viewpoint, we must recognize that there is a role for both private and public institutions. This book examines the market for conservation of natural areas in the USA and the means by which nongovernmental individuals or groups are able, through private transactions, to persuade private landowners to protect wildlife habitat. It considers the efforts of both profit and nonprofit making ventures, analyzes the success of the private conservation market in effectively and efficiently protecting natural areas, and questions how nongovernmental contributions to conservation might be enhanced. The focus is, therefore, not just on identifying the goods and services produced, but also on explaining how they are produced and on analyzing the rules under which producers must operate in both production and sale of goods and services in the conservation market.

Throughout the book, specific examples are provided of strategies designed to overcome imperfections and so maintain an appropriate supply of conserved habitat. Chapter 2 explains conventional economic theory as applied to the protection of natural areas. It identifies how a market for conservation has developed in the USA and explains the characteristics of the market, including the product, the parties involved and the institutional arrangement of the market. Chapter 3 describes the formation of those organizations involved in private protection in the USA. It specifically examines how collective action may enable organizations to capture the nonuse benefits associated with conservation and transfer those benefits from dispersed individuals to private landowners. Chapter 4 explains how various mechanisms are used to protect natural areas while allowing development of conservation products. Chapters 5 and 6 explain how landowners might be encouraged to protect natural areas by deriving income from their use for recreational activities. Chapter 5 examines the role of fee-hunting and Chapter 6, the role of nonconsumptive uses of wildlife. Each chapter addresses the problems of recreational use and makes suggestions for the improvement of existing institutional arrangements. Chapter 7 examines how long-term conservation provisions, traditionally viewed as an encumbrance on the land's title, might be marketed as a valuable asset. It also examines partial residential development as a means of funding the protection of a particular site. Chapter 8 illustrates the importance of appreciating the presence of certain private benefits that are available to landowners who contribute to the conservation of natural areas. It examines two cases where groups of landowners have joined to protect larger areas of natural habitat, independent of any third party organization or government involvement.

Specific case studies are used throughout the book. In each case study, the values associated with protecting the wildlife habitat have been identified and a specific market has been sought that will subsidize the costs of protection in return for certain goods or services. In Chapter 9, conclusions are drawn

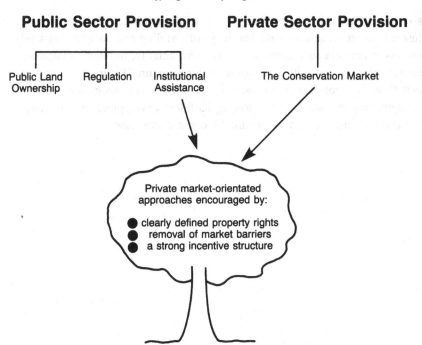

Fig. 1.2. The public/private roles in the conservation market.

about the costs associated with making transactions in the current conservation market. It identifies specific characteristics of the institutional arrangements that are causing transaction costs to remain high and so are preventing the evolution of an efficient market for conservation. Suggestions are made for the improvement of institutional arrangements. The chapter concludes by discussing the future development of the market and warns that private sector conservation may be moving into the realms of public sector regulation.

Many previous books have explained the public sector's role in conservation in terms of direct public provision through land ownership and indirect provision through regulation. This book does not intend to add to that work. Instead, the book concentrates unashamedly on private provision of conservation in the USA through market-orientated approaches. In doing so, however, the book does not attempt to isolate private approaches, and thus reinforce the tendency to treat the public and private sector roles in conservation as a strict dichotomy. With reference to the opening quotation from Leopold: 'There is a clear tendency in American conservation to relegate to government all necessary jobs that private landowners fail to perform,' it is hoped that this book will demonstrate that there is no need to relegate to government all conservation jobs that private landowners fail to perform. Instead, the government might find ways of enabling and encouraging landowners to perform such conservation tasks voluntarily, through an improved conservation market with a strong incentive structure. Figure 1.2 illustrates that there is a role for government in the private conservation market, in strengthening that market, encouraging its future development and in ensuring that a strong incentive structure for conservation exists.

2

The conservation market

Conservation is paved with good intentions which prove to be
futile, or even dangerous, because they are devoid of critical
understanding either of the land, or of economic land-use.

Aldo Leopold (The Land Ethic)

Economic theory and private landowners

Many private landowners have on their property important wildlife habitat
that they may wish to protect, such as wetlands, forests, rivers or lakes. For
much of the recent past, wildlife habitat on private (and public) land has
been destroyed or altered globally, as landowners have sought to exploit the
benefits to be gained from managing the land for other uses. Such uses could
include agriculture, forestry, construction or industrial use (such as landfill
or waste disposal). The private landowner must make a choice regarding the
future use of an area; protection of special features provides only one option.
In assessing how to manage the land, the private landowner will take into
account the costs and benefits of protecting and enhancing the wildlife habitat
against the costs and benefits of alternative land uses.

Today, it is widely recognized that wildlife habitats are valuable resources
when conserved and managed in their natural state. How can we, as a society,
exercise our choice over the future use of such private areas in the interests
of the wildlife resource? Economic theory is concerned with such choices
and the allocation of scarce resources between competing objectives. In order
to appreciate the decision criteria of the private landowner, it is necessary to
understand the characteristics and the distribution of the potential costs and
benefits associated with conserving areas in their natural state.

Neoclassical economics

Economics can be divided into two categories: 'positive' economics and 'normative' economics. Positive economics is often referred to as dealing with
'what is' and therefore tries to explain and predict how people behave. Normative economics deals with 'what should be' and therefore relies upon
underlying value judgments in order to make policy prescriptions. As such,

it is not possible to use economics to decide what is the best solution unless objectives have already been established.

As a discipline, economics combines positive and normative economics in order to understand natural resource problems. In doing so it provides:

hypotheses that can be tested and predictions about behavior (positive economics);
prescriptions from which policy choices can be made (normative economics); and
predictions as to their likely consequences (positive economics).

In order to prescribe policy solutions through economics, the objectives of individuals must be established first. Positive economics makes various assumptions about individuals that help to establish these objectives. In particular, economic theory builds upon the propensity of individuals to act so as to optimize their own interests. In this respect, positive economics is concerned with the personal preferences of individuals. In addition, economics assumes that individuals act in the intelligent pursuit of individual gain with rationality, implying that other modes of behavior are not rational. Such irrational modes of behavior include having regard for others and actions directed toward the public good. Anderson & Leal (1991: 4) considered that 'self-interest may be enlightened to the extent that people are capable of setting aside their own well-being for close relatives and friends or that they may be conditioned by moral principles.'

On identifying humans as rational self-interested individuals, economics concerns itself with theories of how individuals might conduct transactions to optimize their interests. It is through the pursuit of self-interest that outcomes benefiting the public interest are achieved. It is not from benevolence that farmers produce animals and crops to feed a nation, but from their regard to their own self-interest and the goods that they might receive in return for the food. Adam Smith's *Wealth of Nations* acknowledges the presence of an 'invisible hand', which makes it possible to construct all economy based on individual self-interest. He argues that the propensity of humans to exchange goods and services will result in overall benefit to society:

He intends only his own gain, and he is in this, as in many other cases, led by an invisible hand to promote an end which was no part of his intention. Nor is it always the worse for the society that it was no part of it. By pursuing his own interest he frequently promotes that of the society more effectually than when he really intends to promote it.

Smith, 1776: IV.ii

Central to Smith's confidence in a satisfactory exchange taking place is the ability of individuals to have exclusive private rights to economic assets:

the ability to trade assets is dependent upon identifiable, transferable and enforceable property rights. Microeconomic theory examines how individuals pursue such self-interest through determining: (a) how scarce resources are allocated among competing individuals and (b) who benefits from the goods and services provided. It provides important insights into how and why exchanges take place. In this sense, it is useful in helping us to examine the character of conservation and thereby the ability of conservationists to trade in a conservation 'market place'. In particular, three concepts of microeconomic theory have implications in the resource allocation decisions associated with the conservation of natural areas on private land. (For a detailed, coherent discussion of these three concepts and economic theory in general, see Rhoads, 1985.)

Opportunity cost

First, microeconomics has identified that there is always an 'opportunity cost' of using resources. Since resources are scarce, whenever the cost of using resources in one project increases there is a decrease in the resources available for another project, and so a cost in terms of forgone alternatives. An economist's preoccupation with costs might equally be seen as a preoccupation with benefits: a dollar spent on conserving one natural area is a dollar that cannot be spent in protecting the benefits enjoyed from another natural area. It is worth mentioning that as a concept 'opportunity cost' reminds us that the costs relevant to decisions are those connected with future opportunities. Costs that have already been incurred must, in this respect, be regarded as irrelevant; it is the opportunity cost of spending the next dollar in protecting a particular area which concerns us. Nevertheless, decisions regarding resource allocation continue to be made on the premise that too much has already been invested to abandon the project. McLean (1987) referred to this as the *'Concorde* fallacy':

Every year researchers have said, 'Just give us another billion dollars and we will have it airborne within the year.' Every year, they have got their billion dollars. This year, the government is getting restless, but the Minister for Scientific Research says, 'If we don't spend another billion, we shall have wasted the nine billions we have spent so far. Therefore we must spend it.' So they do. The plane gets off the ground; but it never recovers its costs – not even this year's billion.

This raises a second important concept of microeconomic theory – 'marginalism'.

Marginalism

Marginal analysis questions the importance of an additional unit of cost or benefit. When Adam Smith considered his famous paradox of diamonds and water he recognized that 'the things which have the greatest value in use (e.g. water) have frequently little or no value in exchange; and, on the contrary, those which have the greatest value in exchange (e.g. diamonds) have frequently little or no value in use.' (Smith, 1776: I.iv). Smith identified that both goods provided utility, the former because it can be used in a practical way, the latter because the good appeals to our senses. The paradox became known as the paradox of 'value in use versus value in exchange.' Smith appreciated that the higher exchange value of diamonds was partly attributable to their scarcity. However, it was not until the late nineteenth century that the distinction between marginal utility and total utility was made. Rhoads (1985: 25) explained:

The total utility or satisfaction of water exceeds that of diamonds. We would all rather do without diamonds than without water. But almost all of us would prefer to win a prize of a diamond than one of an additional bucket of water. To make this last choice, we ask ourselves not whether diamonds or water gives more satisfaction in total, but whether more of one gives greater additional satisfaction than more of the other.

Answers to marginal utility questions depend upon how much of each particular good we already have. The utility of additional marginal units continues to decrease as we consume more and more. The term 'marginal' is often paired in economics with 'benefit' or 'cost'. 'Marginal benefit' is the additional satisfaction obtained from consuming one extra unit of a good or service. 'Marginal cost' is the cost to produce an additional unit of a good or service.

The concepts 'opportunity cost' and 'marginalism' are inextricably linked but they are not the same; marginalism helps to define opportunity cost but it has wider relevance. For example, when considering the effect of tax incentives on the propensity of landowners to conserve natural areas, we are less concerned with the effect of an increase in the particular tax allowance on the *total* amount of acreage conserved, but more interested in the percentage of any *additional* area conserved as a result of the increase in incentives. This point will become important later when the benefits and costs of conserving natural areas are discussed. This brings us to the third concept of microeconomic theory, which is of importance to policy analysis, that of 'incentives'.

Incentives

Economists since Adam Smith have been seeking ways of using monetary incentives to accomplish public purpose. Schultze (1977) referred to this as 'The Public Use of Private Interest' and commented that: 'Harnessing the "base" motive of material self-interest to promote the common good is perhaps the most important social invention mankind has yet achieved.' His argument is that, rather than exercise political power through regulation to change individual behavior, the incentive structure in which individuals make choices should be understood and then extended towards the common good. The individual's pursuit of self-interest would then help to achieve public good. The importance of this concept will become apparent when we examine the need to provide an appropriate incentive structure for private landowners attempting to conserve natural areas.

The economics of protecting natural areas

The benefits of natural areas

We place a value on natural areas because of the benefits associated with them, much in the same way that we value ordinary goods and services traded in a market place. Most consumer goods and services are 'private' goods: that is, they may only be used by one person at one time. Their use is competitive and so if one person purchases the good, such as a bicycle, then nobody else can buy the same bicycle. Economists refer to this characteristic as 'rival': the bicycle cannot be used by more than one person at a time without subtracting from the enjoyment of another. In addition, it is possible to prevent other people from consuming a private good, for example, the purchaser of the bicycle can prevent other people from consuming it by padlocking it. Economists refer to this characteristic as 'excludable'.

Some private goods can be enjoyed by more than one person at a time, without subtracting from the enjoyment of others. For example, many people could visit an art gallery and view the paintings without affecting each other's enjoyment of the exhibition. Up to a point the exhibition is nonrival, although eventually the number of people entering the gallery will begin to affect the enjoyment of those present, as each struggles to obtain a good view of the paintings. The exhibition is certainly an excludable good, however, since the gallery can exclude the general public and only allow access to ticket holders.

Environmental and welfare economists have paid a considerable amount of attention on commodities known as 'public' or 'collective' goods. A commodity exhibits public goods characteristics when it can be enjoyed by a large number of people simultaneously. A frequently cited example of a public good is a lighthouse. Once lit, many vessels can use the service provided by the lighthouse without in any way affecting the quantity or quality

Table 2.1. *Classification of goods and services*

	Rival	Nonrival
Excludable	Bicycle	Art exhibition
Nonexcludable	Highway	Lighthouse

of warning the light provides for others. In addition, a vessel cannot be excluded from using the lighthouse as it is available for all passing ships. Unlike private goods, public goods and services 'may be enjoyed by all persons simultaneously and the consumption by one person does not subtract from the consumption opportunities of others.' (Weisbrod, 1964). Two essential characteristics of public goods lie implicit in this definition. First, they are seen as nonexcludable: that is, it is impossible to exclude individuals from enjoying the particular good or service. Second, they are seen as nonrival: that is, enjoyment by one person does not detract from the simultaneous enjoyment of another and the consumers are not rivals in their consumption.

Table 2.1 presents classification of goods and services according to these characteristics. Pure private goods are classified in the top left-hand quadrant as 'rival' and 'excludable'. Pure public goods are classified in the bottom right-hand quadrant as 'nonrival and nonexcludable'. Private goods, which can be enjoyed by more than one person simultaneously, are classified in the top right-hand quadrant as excludable but nonrival. Public goods, which are accessible to all but display rival characteristics, are classified in the bottom left-hand quadrant. An example of such a public rival good is a highway, where anyone can pass along the highway, but the number of people using the highway can seriously affect the benefits derived from its use. Anyone who has ever been held up in a traffic jam will appreciate this characteristic!

Traditionally, economists have classified natural areas, along with lighthouses, in the nonrival and nonexcludable category (bottom right-hand quadrant) reserved for pure public goods. The reason for placing natural areas in this category has been the assumption that the area can be enjoyed by many people without diminishing the enjoyment to any one individual and that, indeed, it is impossible to exclude anyone from enjoying the area. Perhaps this assumption has arisen because of the preoccupation in many countries with the scenic qualities of natural areas. Areas such as national parks were often designated as such by virtue of their spectacular scenic beauty. It is only more recently that we have begun to appreciate such natural areas for the recreational and amenity benefits they bestow and for the benefits they hold for wildlife populations, regardless of human presence. In essence, a whole range of benefits stem from natural areas, which suggests that categorizing them on the assumption that they are pure public goods can be misleading in terms of future management. While it is important to recognize that natural areas do provide benefits that are both nonrival and nonexcludable in nature (natural resources do bestow a range of benefits on a large and diffuse population simultaneously), they are also capable of supplying a range of benefits that bear more resemblance to those associated with private goods.

Listed below are some of the benefits that can be attributed to a natural area. While there is general agreement on the relevance of conservation values to decisions over future land use and management of natural areas, there remains a great deal of ambiguity over the relative magnitude of some of the constituent components of total benefits (for discussion, see Gregory *et al.*, 1989; Willis, 1989; Pearce & Turner, 1990; Walsh *et al.*, 1990). The list below is not exhaustive, but serves to illustrate the diverse range of benefits associated with natural areas. The benefits have been categorized into **direct** benefits and **indirect** benefits. The importance of understanding the complex mixture of benefits associated with protection of natural areas will become apparent later in examination of the supply of protected natural areas.

Direct benefits

'Direct benefits' refer to benefits derived from people having access to and directly using the site. They include:

Recreational benefits, derived from the use of the natural area for recreational and amenity pursuits, such as hiking, wildlife viewing, photography or relaxation, and sporting use – the pursuit and taking of wildlife from a natural area.

Extractive benefits, derived from extraction of flora and fauna from natural areas. Extractive benefits can be realized in several ways. For example, many drugs are derived directly from wild plants and even those made synthetically have their origins in plant material. In addition, wild plants are critical for the development of plant-based foods: the creation of disease-resistant genetic types and plants, which will render improved yields, is often reliant on genes from wild plants. Other uses might be made of the resources in the natural area. For example, valuable timber resources might be extracted for furniture-making, plants might be gathered for commercial horticultural production and animals hunted for food (see Prescott-Allen & Prescott-Allen, 1986). In each case, the extraction of a particular resource is not necessarily incompatible with conservation of the habitat and as long as extraction is controlled over time, it need not necessarily destroy nor damage the natural area.

Educational benefits, derived from the study of natural areas, providing information on, for example, evolutionary processes and the nature of ecosystems (see Allen, 1988).

Cultural benefits, derived from natural areas, particularly when a culture has a strong tradition of associating natural resources with spiritual beliefs.

Indirect benefits

There are several different types of indirect benefit associated with natural areas. They are termed 'indirect' benefits because, by their very nature, they do not require the beneficiary to enter onto or use the natural area in any direct manner.

Natural functional benefits are derived from the role that natural areas play in assisting in the effective functioning of the ecosystem, such as watershed protection, photosynthesis, regulation of climate and soil production; storage and cycling of essential nutrients; absorption and breakdown of pollutants; and environmental monitoring (see, for example, Newman & Schereiber, 1984; Myers, 1988).

Aesthetic benefits are derived from the sense of well-being that people experience from being able to view the natural area. Such benefits can be enjoyed from an adjacent area of land and do not necessitate access to the natural area itself.

Existence benefits are derived from the knowledge that a piece of land exists, regardless of whether the beneficiary is able to see it. McNeely (1988: 24) commented that existence benefits reflect 'the sympathy, responsibility, and concern that some people may feel toward species and ecosystems'.

Option benefits are the benefits derived from being able to use the resource at some time in the future. For example, an individual may not wish or be able to visit a natural area at the present time, but would like the forest to be conserved so that he or she may have the option to use it in the future. Interestingly, it has been noted that the benefits derived from the option to visit natural areas in the future may encompass hope, opportunity, dreams and fellowship that are independent of the values of actually visiting the place (Cicchetti & Freeman, 1971). Alternatively, the natural area may contain resources for which we have no use at the present but which we may wish to harvest in the future. In this respect, option benefits acknowledge the need for an 'ark' of diversity, which may be needed in the event of future biological, socio-economic or political events.

Quasi-option benefits are derived from the uncertainty regarding future availability of information associated with resources that might otherwise face irreversible damage. Conrad (1980) summarized quasi-option value as 'the expected value of information gained from delaying an irreversible decision'. The use of plant material for the manufacture of drugs illustrates the importance of recognizing quasi-option benefits. Pearce & Turner (1990) pointed out that 'the issue at stake is what value has yet to accrue from plants which *are* endangered but which have yet to be "screened" for medicinal properties. Brief appraisals exist for only one plant in ten. Detailed assessments exist for only one in a hundred.'

Bequest benefits are derived from the knowledge that a particular area will be conserved for enjoyment by future generations. The value placed on the future use of the area links bequest benefits very closely to existence, option and quasi-option benefits. The presence of bequest benefits is often used to economically justify conservation of natural areas (see, for example, Krutilla & Fisher, 1975).

Supply of conservation

The importance of the nature and range of these 'economic' benefits is associated with the effect that they have on conservation. It is already apparent that the benefits derived from a natural area will not be exclusively enjoyed by the person who holds the property rights to the land. The extent and range of benefits available involves a much wider population of potential beneficiaries simultaneously. In this respect, they often display the nonexcludable qualities of a 'public good', while at the same time displaying the types of benefit associated with a 'private good'. In fact, as has been demonstrated, conserved natural areas reveal the characteristics of both types of good – they can bestow a range of benefits simultaneously on a large and diffuse population. Equally, it is possible to exclude people from full enjoyment of nature conservation if that enjoyment involves access to and use of the particular site.

In categorizing natural areas as pure public goods, there is a tendency for economists to become preoccupied with the scenic benefits of an area, attempting to classify natural areas per se, rather than the goods and services derived from them. Table 2.1 can be revised to accommodate the different types of goods and services derived from natural areas. In addition, the table

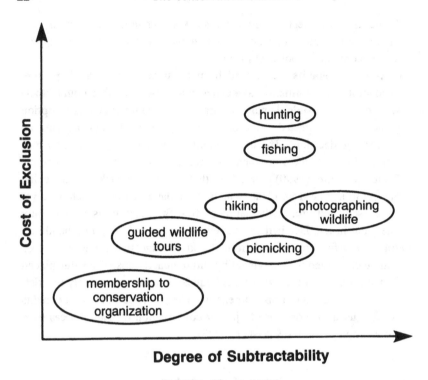

Degree of Subtractability

Fig. 2.1. Classification of conservation products.

is revised to suggest that few goods and services can be judged as purely 'rival' or 'excludable', but must be considered in terms of how costly it is to exclude individuals ('excludability') and the extent to which one person's enjoyment subtracts from another's ('subtractability'). Thus, the revised illustration (Fig. 2.1) enables us to recognize that the problems encountered in managing land are not as clear cut as the original categories suggest: we can exclude people from a site, but it might take a good deal of fencing and/or 'guards' to enforce such exclusion. Equally so, more than one individual can enjoy a particular area, but eventually the extent to which each enjoys being there will be affected by the number of other people present. The matrix divisions of the table have been removed to demonstrate that excludability and subtractability are continuums, rather than discrete entities. For example, the hunting of an area by one person subtracts from the enjoyment of another individual hunting the same area. If the hunting is on private land then either person can be excluded. However, the degree to which this particular 'good' is subtractive and the extent of the cost incurred in excluding other hunters will depend on a host of other variables, most particularly the size of the site, but also factors such as the behavior of each individual hunter and the extent to which they will conform with rules without the need for expensive enforcement.

While not providing perfect and conclusive categorization, the classification of conservation goods and services in this way can help to alert us to the problems of managing natural areas and supplying such goods. The 'subtractability' classification warns us that there is often a human 'carrying capacity' for a particular activity on a particular area of land. The need for exclusion raises the question of the best means of achieving exclusion from a particular site and, as discussed late, this need not necessarily involve fencing an area.

Probably more important, in terms of appreciating the problems of supplying conservation products and managing the natural areas from which they are drawn is the realization that more than one product can be produced from a single site. Indeed, production need not be limited to conservation products, but may include several other closely related and compatible products. The classic example is forestry. A well managed forest can support a selective logging regime producing timber, while also providing for a whole host of recreational activities and the indirect benefits identified above (natural functional, aesthetic, option, quasi-option, existence, bequest). Equally, agricultural land can produce traditional agricultural products while at the same time producing conservation goods and services associated with the plant and wildlife populations supported by the land.

In understanding the evaluation, supply and financing of conservation products, we must appreciate that natural areas can support multiple goods and services. With careful management some goods are capable of being enjoyed simultaneously by more than one individual: while others must be enjoyed individually. Examining the benefits to be derived from a natural area provides only one side of the conservation picture. In order to understand the relevance of multiple supply, it is important to understand the costs associated with conservation of natural areas. The key questions in predicting how much wildlife habitat will be conserved are: (a) what is the cost of conserving the land in its natural state? (b) who will bear that cost?

The costs of protecting natural areas

The costs of conserving natural areas are often large (Willis *et al.*, 1988). While many people might enjoy benefits derived from a particular area of wildlife habitat on private land, it is the private landowner who is likely to bear the majority of the total costs of protection. There are three elements of cost involved in such protection. First, there is the cost of forgoing the right to use the land without restriction, such as the right to log a forest, sell the timber and replant or cultivate the land for agricultural use. This may be referred to as the 'opportunity cost' of protection. Although the majority of this cost is incurred by the private landowner, this may well be supplemented by a secondary cost to society (such as reduced economic growth) as a consequence of the modified management practices. It is generally acknowledged that the 'opportunity cost' is the most significant cost incurred. Second, if a guarantee of protection is sought, there is the cost of negotiating some type of formal protection agreement for the land, specifying the rights and duties of parties to the transaction, usually through some legal mechanism. If perpetual protection is sought, which would be necessary to ensure some of the benefits suggested above, there must be some perpetual body that will oversee the enforcement of the protection agreement. The cost of monitoring and enforcing the agreement is the third cost of protection.

If landowners are going to bear the cost of protecting wildlife habitat in its natural state, then we could expect that the benefits they receive in return must be at least equal to their costs. This is consistent with the economist's view that the landowner will act in his or her own interest. Figure 2.2 is a marginal benefit/marginal cost diagram illustrating how a private landowner might decide how much natural area to conserve. Positive economics uses the concept of marginal benefits and costs to predict a producer's behavior when deciding how much of a particular commodity to produce. The assumption is that the producer will carry on producing to the point at which the marginal cost of increasing production by one unit is equal to the marginal benefit derived from one extra unit of production. It is at this point that the producer accepts that maximum net benefits of production have been obtained. If the landowner takes into account his or her own benefits, then he or she will protect 'Ql' amount of land, that is the point at which the marginal benefits of protection (the curve 'B − landowner') are equal to the marginal costs of protection (the curve 'C − total'). However, we have already witnessed that a whole host of other people derive a range of benefits from the land being conserved in its natural state. They are represented by the curve 'B − all others'. The point at which they would like to see natural areas protected is 'Qao'. The answer to ensuring that the amount of habitat protected suits the requirements of 'all the other' people involved seems quite simple. If they were to bear the additional cost, then the landowner might increase protection to a level 'Qao'. The task facing the landowner is to find

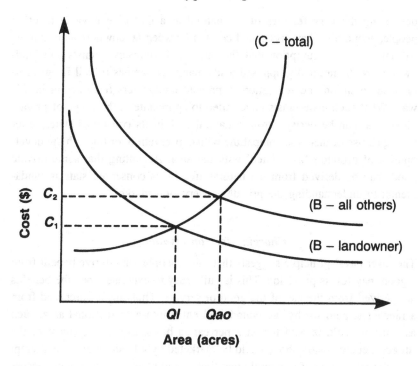

Fig. 2.2. Decision-making: the costs and benefits of protecting natural areas on private land. *Ql*, amount of land the landowner protects; *Qao*, amount of land 'all other' people demand to be protected; C_1, cost of protecting *Ql* amount of land; C_2, cost of protecting *Qao* amount of land.

ways of making all of the other people contribute to the *extra* cost of the conservation (C_2-C_1).

Recovering the costs

Often it is assumed that the only private commodities that can be derived from the natural area are those which will dramatically alter the area itself, for example extraction of resources from the area (timber, wildlife, plants) for sale as primary products in some production process. By virtue of the ability to exclude all those who do not pay, such private products can be easily sold to provide the landowner with some form of income and return on the capital value of the land. However, as a consequence of this assumption, predictions concerning the future use of the natural area have been fairly pessimistic. Recognizing that the landowner can derive income only by destroying the very features of the natural area that give it value to other people, economists have predicted conflict between landowners and 'conservationists'. The assumption that the interests of conservationists and landowners are diametrically opposed leads many economists to call for government intervention and regulation to prevent landowners from acting in this way. What such economists have failed to appreciate is the range of private goods that can be derived from a natural area in its conserved state, either through a use of that area compatible with conservation, or through the development of products 'associated' with the area. Accepting that some private goods can be derived from the natural area in its conserved state is fundamental to understanding the private conservation market.

Charging for conservation

The 'user pays' principle suggests that those people who derive benefit from a good, pay for its provision. This is quite easy to exercise when the benefits are derived from the use of the good or service. Thus, meat purchased from a rancher is paid for by the pound, and entry into a recreational area, such as a theme park, is paid for on a per capita basis. Equally so, most of the 'direct' benefits listed above could be marketed by a landowner in an attempt to realize the value of a natural area: timber and plants could be sold; access could be granted for recreation and sport at a charge; licenses could be granted to scientists wishing to use the area for research investigation. However, as shown above, many of the benefits associated with conservation are 'indirect' benefits. Such 'indirect' conservation benefits have certain characteristics that affect the way in which they can be harnessed by the landowner and subsequently exchanged for payment. First, the private landowner may not necessarily be able to exclude a person from deriving benefits from the presence of a conserved area. For example, although access may be denied, aesthetic benefits may be enjoyed from a neighboring piece of land or nearby roadside. Existence, option, quasi-option and bequest benefits cannot, by their very nature, be excludable. Functional benefits will be awarded to a vast range of people, not only those people within the immediate proximity of the particular site (e.g. benefiting from watershed protection), but also people from around the world (e.g. benefiting from the effects of climatic regulation). The diffuse nature of the beneficiaries of natural areas explains why habitat loss in one region, country or continent is of concern in another. However, in terms of charging for conservation, it presents landowners with a major problem – the 'free-rider'. If free-riders cannot be excluded from enjoying an area of land, then individuals will realize that they need not pay for the product in order to enjoy it, but can 'free-ride' on other's payments. Second, there is the nonrival nature of most indirect benefits. For example, one person's contentment in knowing that a particular wildlife habitat is protected may exist without impinging on another person's benefits. Third, conservation benefits are widely and thinly distributed over a large population. Like any product, the size of the conservation benefits sought by individuals varies according to their own needs and wants. Landowners wishing to sell conservation have to find ways of capturing these diffuse benefits.

The conservation market

In their search to share the costs of protecting natural areas for wildlife, landowners in the USA have arrived at many innovative means of charging for conservation, principally through the development of property rights under market principles. Whether a landowner seeks income from the natural area to make a profit or merely to cover costs, the 'invisible hand' of the market directs individuals to provide protection for natural areas, without the need for government intervention. Income-generating ventures include charging for consumptive use of wildlife, such as fee-hunting, and for nonconsumptive use, such as access to view, photograph and interpret wildlife. Other projects seek to capitalize on the benefits associated with natural habitat, such as developing residential property adjacent to the natural area and using the environmental quality of the area as an additional marketing tool. Each venture uses the allocation of property rights to enable the landowner to market a product associated with conservation, while protecting the natural area itself. A sophisticated range of property rights has been developed in order to allow successful transactions to take place, while continual improvement of the market helps to overcome any barriers to trading.

It is important to appreciate how the market has developed and to recognize that there are problems in developing marketable goods and services from the protection of natural areas. The conservation market differs fundamentally from other markets in which goods and services are traded. Economists will refer to a 'perfect market' as one in which there are many buyers and sellers, a homogeneous product, perfect knowledge of transactions by the people involved in the market and an ability for each supplier to affect the price of a good. It is quite possible that there is no real market place that fulfills *all* of these conditions. The conservation market, however, has a large number of imperfections, of which the following are regarded as the most important.

1. **No single product.** While most markets deal with the transactions of fairly homogeneous products, the conservation market has a central goal

(the protection and management of natural habitat), but achieves this end through the sale of several different products, employing a whole host of property rights.

2. **No central market place.** Transactions take place in a collection of related markets, the specific market being dependent upon the type of product being offered and its location.

3. **Fixed supply.** Subject to certain physical and legal constraints, the uses that might be made of any one area of natural habitat can change over time. However, the amount of wildlife habitat existing in the USA is relatively fixed, with only small areas capable of being added to the whole stock by virtue of habitat restoration.

4. **Lack of knowledge.** The lack of a central market place means that there is a lack of knowledge regarding conservation transactions. This is also due to the desire on the part of many landowners and conservationists to keep transactions secret, particularly when they involve only two or three parties. Traditional dissociation of wildlife and markets means that there is still a reluctance to reveal that we are *dealing in diversity*. When conservation is openly marketed to a large number of individuals (such as a recreational enterprise), then good information is generally available on the price of purchasing a right to a natural area. In both cases, there remains the need for further analysis of the costs of protecting and managing natural areas.

5. **Complex tools of transaction.** Unlike most other markets, transactions in the conservation market are often dependent upon complex agreements, involving more than two parties. Agreements need to ensure protection of an area of natural habitat (the end goal), while at the same time charging for some type of associated good or service (the product). This means that apart from the two normal parties to the transaction – the supplier and the customer – there has to be a third party who will, for example, oversee the protection of the site in perpetuity. Inevitably the time taken to complete a deal may be lengthy and may cause considerable expense.

In the development of the private conservation market in the USA, many innovative solutions have been adopted in order to improve the efficiency and effectiveness of operation. Work carried out to improve the market might be categorized into three areas:

1. definition and development of products;
2. establishment and development of agents to transactions; and
3. the incentive structure for conservation.

Development of conservation products

From the forgoing discussion, it is evident that several tasks face American landowners if they are to develop conservation products. They must:

1. find ways of excluding people from the product;
2. assess the price that people are willing to pay to obtain the product; and
3. target the marketing in accordance with the perceived customers.

Physical constraints on specific areas of land will, to a large extent, dictate the type of goods and services that are compatible with protection of wildlife habitat and are capable of being provided. Excluding people from the product does not necessarily involve fencing off an area, putting in a turnstile and charging an admission fee. Restricting access is certainly one option: careful zoning and management of areas can make access provisions compatible with conservation of natural habitats. However, other mechanisms have been sought in the USA to achieve exclusion.

One way of compensating for the nonexcludable nature of the benefits derived from a conserved area is to offer associated products that are readily made exclusive to those who subscribe to the area's protection and management. These take the form of: information about the area, such as leaflets and newsletters; guided tours of the area; membership of a supporter's club, including certificates, badges and car-stickers; and/or voting rights over the management practice of the area. Whatever personal needs motivate individuals to contribute to a common cause, personal satisfaction is undoubtedly gained from recognition of the individual's contribution. Successful landowners and organizations have found appropriate ways of acknowledging this contribution and reinforcing the individual's behavior.

It has been stressed that conservation benefits are often nonuse benefits from which it is difficult to exclude people. Conservation organizations must make full use of this characteristic in their product development. People are willing to contribute to conservation without actually visiting the site. This has been evidenced over the years by the amount of funds the public are willing to donate to the preservation of Antarctica, despite that fact that individual contributors are unlikely ever to visit this last wilderness. A whole host of conservation schemes is now open to customers, from 'adopt-a-tree' by mail order, to 'Give As You Earn' income tax arrangements. The task set is to find ways of selling the conserved area, without necessarily selling property rights to the land itself.

In some cases where there are management advantages in having an active group involved in the site, then some of the property rights may be transferred. For example, an area may be leased to a local conservation group made up of volunteers willing to carry out specified operations on the site, such as weed clearance or tree planting. In such cases the landowner is not only selling the conserved area for the price of the lease, but also for the labor and skills offered by the group. As a result of the group's involvement, contributions may be made from skilled and experienced botanists, biologists and ecologists. Like the leasing of commercial property, tenant selection can be very important. Conservation organizations can be particularly important in providing an agency for arrangements between landowners and interested groups.

The establishment of conservation agents

There is widespread recognition that the private sector has had an important role to play in the conservation of America's natural resources: 'Americans have learned not to wait for Washington to act, we have developed a system to empower, structure, organize at the grass roots so that people can act.' (Scott, 1988). A good deal of the attention, which has been placed on this private sector provision of conservation, can be attributed to the work of the conservation organizations. Local, national and international, nonprofit organisations have played an important role in the development of the conservation market in America. Essentially, they have acted as agents bringing together the wishes of many thousands of individuals and using their collective strength to achieve protection of natural areas. (Chapter 3 specifically examines the work of these nonprofit organizations and their role in the conservation market.)

In such instances, instead of a request to support individual natural areas, the customer is asked to contribute to an organization that undertakes to carry out work consistent with the customer's objectives. This is the classic case of the nonprofit-making conservation organization. Rather than sell access to or use of a particular site, a conservation organization will sell a number of less tangible products, such as membership, a newsletter, a badge and the knowledge that the contributor is helping to save America's natural areas.

Conservation organizations comprise a type of 'club', relying on the fact that individual members will appreciate the mutual benefits to be derived from sharing the production costs of conservation in return for consumption benefits (Buchanan, 1965). However, it has been suggested that larger 'clubs', such as national conservation organizations, will need to provide their members with some sort of excludable benefits in order to retain their interest (Olson, 1965; Berglas, 1976). In this respect, providing products associated with their work helps conservation organizations to reinforce the belief that contributors are receiving benefits from which noncontributors are excluded.

Actors in the conservation market

Throughout the book, all parties present in the conservation market are discussed and their specific role in the market analyzed. In all, four groups of decision-makers can be identified as relevant in exploiting the opportunities of protecting natural areas. These are:

1. the landowners, who hold the key to resource use and protection through their ownership of property rights;
2. the public, who demand a vast array of goods and services associated with the natural environment, each with its own values;
3. public decision-makers, who both directly and indirectly affect the incentive structure of the market through the design of public policy; and
4. the private conservation organizations, which act as agents between the landowners and the public and which seek to improve the institutional arrangement of the market by lobbying public decision-makers for new policies.

Figure 2.3 portrays the various roles of each group: the landowner with his/her resource; the public who demand all sorts of benefits from the resource; the conservation organization, which can act as a valuable agent satisfying public demand by negotiating with landowners; and finally, the public decision-maker who, to a large extent, determines the trading environment.

The incentive structure for conservation

Throughout this chapter there has been an implication that we can persuade private landowners to protect natural areas by contributing to the costs of conservation through the purchase of goods and services. This implies that the only incentive that the landowner receives is the cash paid for the particular commodity. In fact the landowner may receive other types of benefits, which act as an incentive to augment conservation action. Some such benefits are supplied by the government in the form of tax subsidies for the landowner. These are discussed in detail in Chapter 4. Why, in these private transactions between conservation-minded individuals or organizations and private landowners, should the government intervene by providing additional

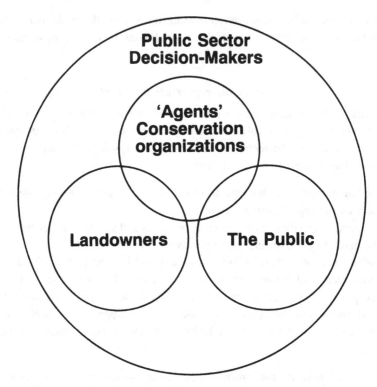

Fig. 2.3. Actors in the conservation market.

incentives? The answer to this question lies in the aforementioned 'free-rider'.

In examining the benefits associated with protecting natural areas, the existence of the nonexcludable and nonrival characteristics commonly associated with public goods has been revealed. The presence of such characteristics in the total benefits derived from the natural area suggests that people cannot be excluded from enjoying the area and, indeed, may do so without ever visiting the area. The inability to exclude people from enjoying a good leads to an inability to charge them directly for the amount of the good that they consume. Collecting revenue through personal taxes is generally thought of as an efficient means of levying a charge for public goods. The revenue collected can be spent on conservation and the cost of protecting a natural area is distributed amongst all tax payers. It is a way in which society can eliminate the 'free-riders'.

Acknowledging the presence of free-riders helps us to appreciate the need for some sort of government intervention in conservation. The government has the ability through tax legislation to ensure that no one is allowed to avoid contributing to the costs of conservation. However, while the presence of free-riders suggests that the government might be best placed to collect revenue for conservation, it does not necessarily justify public provision of conservation. This book suggests that, in many cases, the most cost-effective means of the government supporting conservation is through the encouragement of private provision of conservation. According to microeconomic theory, landowners can be expected to act in their own self-interest and manage their land in a way that produces the maximum net benefits for themselves. However, it is possible that landowners will choose to protect natural areas if the correct form of incentive is present.

Return to economic theory

Marginalism

This chapter began by exploring three concepts of economic theory that are useful in understanding the protection of natural areas on private land: (a) opportunity costs; (b) marginalism; and (c) incentives. It is worth, at this point, expanding the concept of marginalism and, in particular, its limitations when applied to natural area protection. Figure 2.2 explained that society needed to provide the landowner with some form of incentive to persuade him or her to increase the amount of natural area protected to a point Qao. The diagram is useful in explaining how the private landowner's decision on whether to conserve the land's natural features is based on the costs and benefits of conservation. However, if we wish to fully understand natural area protection on private land, we must accept that the area which has been identified as supporting valuable plant and animal life and in need of protection is likely to be of a fixed acreage.

Figure 2.4 illustrates a case where a landowner of such an area is deciding whether to protect the land in its natural state. According to the landowner's marginal costs and benefits of protection, he or she is likely to protect an area of acreage Ql. In Figure 2.2, it was pointed out that society would have to purchase enough conservation to reduce the landowner's costs of protection by $C_2–C_1$ and so ensure that an area Qao (the amount 'all others' demanded) was protected. It is at this point that marginal economics lets us down; for private land comes in fixed supply units.

Suppose that the area of land owned by the landowner is 100 000 acres: there are two problems with protecting to a point Qao. First, the animal and plant life may depend on the protection of the entire site. Protecting the area

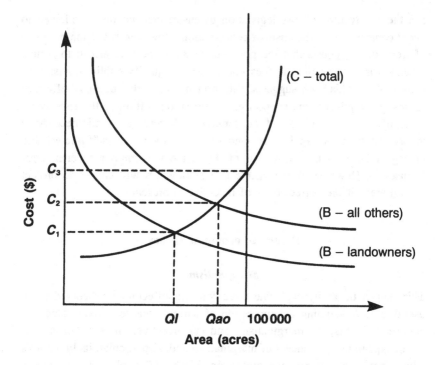

Fig. 2.4. Decision-making: the costs and benefit of protecting natural areas on private land (100 000 acres). *Symbols*: see Fig. 2.2, p.25.

Qao acres may not be sufficient to ensure that the wildlife on the site is supported. This will be particularly true if special features that are vital for the wildlife's existence (such as food, water, nesting, grounds, etc.) exist on the remaining acreage. Second, while wildlife is not inclined to deal in 'marginals', neither are landowners. The issue of whether to protect the land may demand an all-or-nothing decision. This is particularly true when land-owners are tempted to sell land for development.

The conservationists will eventually appreciate that they must protect the entire 100 000 acres or lose the site and that tinkering around at the margins is insignificant in this case. The fact that protection of natural areas does not conform with 'marginalism' in this way means that society may have to revise its ideas about the benefits it might receive from protecting a particular site and acknowledge the all-or-nothing situation. In Figure 2.4, society will have to step up its purchase of conservation to C_3–C_1 in order to ensure protection of the 100 000 acre site. In doing so, however, it will have to examine the benefits it will lose by not being able to protect other sites and it must consider the 'opportunity cost' of the transaction.

The need for flexible institutions

The constraints imposed by the fixed supply of conservation land explored above can, to some extent, be overcome by the conservation market. In the case where a conservation agency is forced to purchase an area of land greater than the optimal acreage, then it might be possible to sell on the remaining land once protection has been secured. Indeed, Chapter 7 explains how the organization may recover all of its costs of protection (the purchase price of the land and any costs of acquisition) by developing the additional acreage. The ability of a conservation organization to act in this cost-efficient manner is very much dependent upon the institutional structure of the market. For example, having a diverse range of property rights at its disposal and the ability to raise finance quickly may be crucial to the organization in protecting the threatened land.

Conclusion: improving the operation of the market

In the market for conservation, transactions are never simple. As already explained, consumers are widely dispersed, with no central market place to bring together the buyers and sellers of conservation products. Indeed, the products themselves are diverse in nature and a complex range of tools is needed to exact each transaction. It is no surprise, therefore, that players in the conservation market have concentrated their efforts on finding means of improving the operation of the market so that more transactions might be performed more effectively and at a reduced cost. They have, in effect, sought ways of improving the 'efficiency' of the market, so that more resources (financial and otherwise) can be spent on conservation and less on performing the transactions that enable conservation to take place. In short, they are aware of the 'opportunity costs' of conducting each transaction. Economists refer to the costs of trading in the market place as 'transaction costs'. Eggertsson (1990) defined transaction costs as 'the costs that arise when individuals exchange rights to economic assets and enforce their exclusive rights.' The assumption that there are no barriers to exchange is an unrealistic one: in practice the costs of undertaking a transaction can be sufficient to prevent the exchange being completed. Some economists argue that an underprovision of protected natural areas might stem not from a 'failure' of the private conservation market but from the obstruction to market trading of high transaction costs.

Recognition of transaction costs in market trading confirms that the nature of the market and the 'institutions' that govern its operation influence the amount and nature of transactions taking place. Ostrom (1986) defined institutions as sets of working rules (formal or informal) that are used to determine:

who is eligible to make decisions in the market place;
what procedures must be followed;
what information must be provided; and
what benefits individuals will reap according to their actions.

The institutional arrangement of the market has a significant influence over market effectiveness and efficiency by:

1. affecting the magnitude and distribution of the costs and benefits of protection (functions of the institutional structure);
2. establishing rules and providing incentives that influence decisions of individuals and organizations (by providing opportunities for them to act on the values they associate with protected wildlife habitat);
3. determining the supply of useful information on the opportunity cost of alternative land uses; and
4. allowing an efficient transfer of benefits from the public to the private landowner, enabling an optimal amount of land to be protected.

Institutions for protecting land should, therefore, create the right incentives to allow private landowners to maximize their own benefits of protection.

In the light of technological, social and economic changes, governments have an important role to play in the conservation market. Governments not only provide the mechanisms for formal protection, in the form of defined and enforceable property rights, but also influence the incentives and choices available to individuals who may wish to protect wildlife habitat through their interaction in the conservation market. A constraint on the range of choices available or the absence of incentives presents a rigidity which can limit opportunities for reducing the costs of conservation. Any institutional rigidity present may constrain individual or organizational initiative and so restrict the amount of land or the efficiency with which land is being protected. Better institutional design with greater flexibility may result in less need for public funding, either because more financial resources are obtained from the private sector or because the same level of supply can be provided at a lower cost if the private interest in protection can be effectively harnessed (Schultze, 1977).

Government intervention in the conservation market may improve the efficient operation of the market through the introduction of changes to the institutions. There are various institutional devices that might be employed to encourage landowners to protect natural areas: some will be specifically designed to do so, others may not. Government intervention could also prove useful in identifying and eliminating those characteristics of any market situation that act as disincentives to transactions. Throughout the remainder of this book a more in-depth analysis of the institutional arrangement of the market includes: (a) an examination of private conservation organizations acting as agents for the customers or suppliers; (b) the legal and other mechanisms used to protect areas; and (c) the incentive structure of the market. The book explains how the private sector has developed the conservation market in order to overcome some of the costs commonly associated with transactions enabling the protection of private land. However, in doing so, it reveals where high transaction costs still exist and are inhibiting efficient private provision of natural area protection. Suggestions are made as to how transaction costs might be lowered further in order to improve the efficiency of the conservation market and recommendations are made for institutional reform.

Some economists believe that if property rights are well defined and transferable, institutional change will occur through market transfers. They argue that whenever benefits exceed the cost of changing institutional arrangements, the ever-faithful 'invisible hand' of the market will transfer rights to those positions where they produce most benefits. Other economists recognize that there is a role for government in altering the institutional arrangement of the market. Analysis of the conservation market will, therefore, identify instances where government intervention is needed to exact institutional reform and instances where market forces should prevail.

3

Collective action

The Nature Conservancy, like any organization, is simply a
group of people working together to do something they cannot
do individually.

The Nature Conservancy, 1955

Introduction

People often combine as a group to purchase goods that they find difficult
to obtain as individuals. By combining effort, the group is able to reduce the
transaction costs commonly associated with the purchase of 'conservation',
such as the collection of funds, the collection of information, the design of
protection mechanisms and the monitoring and enforcement of the protection
mechanisms.

The organizational development of the conservation movement in America
began in the latter part of the nineteenth century. At first, organizations were
made up of small numbers of individuals, often from academic backgrounds.
(For further reading of the development of the conservation movement in
America, see Fox, 1981; Trefethen, 1975; Dunlap, 1988.) The conservation
movement has since grown into a large collection of organizations, national
and local, with vast memberships. It is estimated that the revenues of the
dozen largest environmental groups in the USA currently amount to around
$250 million per annum (*Wall Street Journal*, 1990). The success of each
organization lies in the unique style with which it approaches conservation.
Several of the larger conservation organizations have focussed on the direct
private provision of wildlife habitat, while others have concentrated on lobbying government for increased public provision. This chapter looks at ways
in which the former work with private landowners to encourage and enable
habitat protection on private land. Two organizations are examined as case
studies:

The Nature Conservancy – a large, international organization, founded in
the USA in 1951; and
Land trusts – small, local organizations, established all over the USA.

The Nature Conservancy

The Seventeenth Annual Report from the Council on Environmental Quality referred to The Nature Conservancy as 'arguably the most atypical, innovative, result orientated private sector conservation group today' (Council on Environmental Quality, 1986). The Nature Conservancy was founded in 1951, although its roots go back to 1915. Its mission is 'to preserve plants, animals and natural communities that represent the diversity of life on Earth by protecting the lands and water they need to survive.' The Conservancy has more than 740 000 members and has protected almost 8 million acres in the United States and Canada, through the purchase of private land or negotiation of agreements with private landowners (The Nature Conservancy, personal communication 1994). Between 1984 and 1986, the Conservancy's expenditure on endangered species was $180 million; three times that of the federal government during the same period. The Conservancy holds approximately 1500 preserves, covering over 1 million acres worldwide and representing 1750 rare species and natural communities: the largest private system of nature sanctuaries in the world. The Conservancy adopts a nonconfrontational approach to conservation whereby it seeks to lead by example, to influence the management of public land and to increase government funding of conservation of natural areas. To do so it must have its own house in order and, therefore, can be seen to strive for the protection of significant areas and the implementation of sound management practices.

Organization and funding

The Nature Conservancy has 'Chapters' in each of the 50 American States. Its national headquarters are located in Arlington, Virginia. It portrays itself as a large, international organization, with 40 years of experience and a collection of many skilled people, yet one with strong local roots through its state chapters. The extent of local involvement in the Conservancy's activities would suggest that it is successful in this respect.

The Conservancy's ability to reserve most of its funds for the direct protection of private land has been one of its most attractive features for some time. In the fiscal year 1993, 85% of its funds were allocated to program expenses and capital allocations (see Fig. 3.1). In 1993, the Conservancy received more than $122 million in the form of subscriptions, contributions and grants: about one third of the private funding donated to national environmental groups in the USA. The majority of the Conservancy's funds, in terms of donations and land contributions, are derived from individuals (73% in

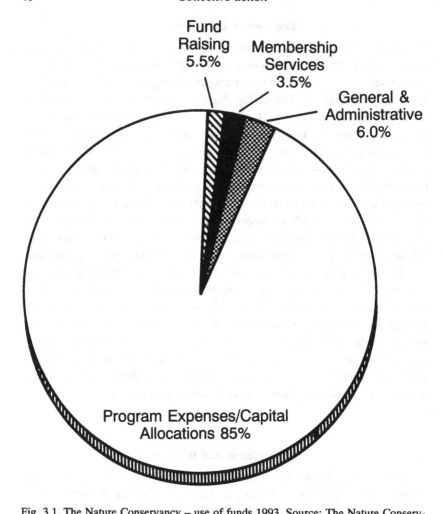

Fig. 3.1. The Nature Conservancy – use of funds 1993. Source: The Nature Conservancy (1993).

1993), with corporations contributing 11.5% of funds and charitable foundations and others 15.5% (Fig. 3.2). Surveys of corporate contributions to the Conservancy suggest that the size of a corporation's donation increases in accordance with the size of its advertising budget and its cost of meeting environmental regulations (Griffith & Knoeber, 1986). The Conservancy accepts gifts of cash, real estate and other valuable chattels. All attract full income tax or corporate tax deductions, since the Conservancy is registered as a nonprofit, tax-exempt organization. Gifts of real estate attract additional tax benefits (see Chapter 4).

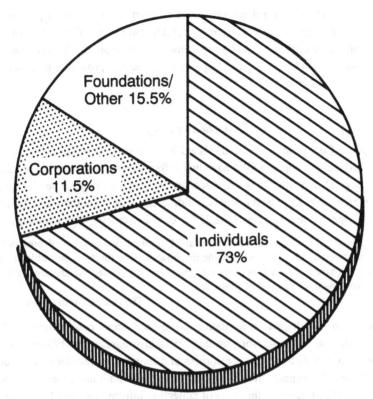

Fig. 3.2. The Nature Conservancy – source of funds 1993. Source: The Nature Conservancy (1993).

The Conservancy has a policy whereby funding raised in a particular state remains in that state. The rationale behind the policy was to strengthen each chapter's identity and to encourage local citizens to contribute to the Conservancy because they perceived that their funds were furthering the protection of local lands. This policy, however, has resulted in a regional inequality between Conservancy activity and biological diversity in certain states. A chapter may have problems raising funds in a state with a small population or where local citizens are unwilling or unable to contribute to the Conservancy. States that are rich in biological diversity but poor in funds include Utah, Georgia, Nevada, Alabama, Tennessee and North and South Dakota. Typical of its positive style of management, the Nature Conservancy has begun to address this problem. In its conservation strategy (The Nature Conservancy, 1990), the Conservancy recognized that 'distribution of our spending is still not well correlated with most measures of biodiversity.' Through a General Fund, in which donations made without specification of particular allocation are deposited, and separate capital campaigns, the Conservancy has begun to distribute central funds amongst its chapters. In addition, it intends to encourage those state chapters with wealthy programs to generate resources for other states.

Protection strategy

The Nature Conservancy was established by a group of biological scientists, which probably explains why it has always focussed its resources on protecting biological diversity. It is often recognized that any land protection program which seeks to preserve biological diversity requires a thorough natural area and species inventory as a basis for decision-making and planning:

Nothing is more fundamental to The Nature Conservancy's mission of preserving biological diversity than precise scientific information. Our ability to protect land and to manage it wisely depends on a sound system for identifying the species, habitat, and ecosystems most in need of protection.

The Nature Conservancy, 1989

Collection of adequate information on natural areas can be one of the most prominent obstacles for any conservation organization. Without such information it is difficult to set priorities and monitor achievement of objectives. When the organization's objectives include national or even global protection of rare species, the costs of collecting information may be substantial. The Nature Conservancy has tackled this problem by initiating a national inventory system.

The Natural Heritage Inventory Program is a network of natural area inventories designed by the Conservancy and established with assistance from the Conservancy, but administered by public agencies (such as the 'department of natural resources') or other institutions in each state. The creation of a national network of state-by-state inventories began in 1974. The Conservancy has fully supported the Program since its inception. Since 1989, all states in the Union have participated in the inventory (Alaska and Alabama being the last to join).

The Heritage Program employs over 400 scientists and information managers to collect, organize and disseminate information. It has enabled a comprehensive directory of rare species and ecosystems to be compiled by maintaining an ongoing inventory of plants, animals and natural communities threatened on a statewide or national basis. Work is carried out through literature searches (recording observations made in unpublished and published documents) or field work. Species are graded according to whether they are:

1. federally listed as endangered or threatened; or
2. suspected or known to be extinct.

Each species is then classified:
1. **critically imperilled** (found in no more than five locales, or numbering no more than 1000 individuals)
2. **imperilled** (found in no more than 20 locales, or numbering no more than 3000 individuals); or
3. **rare** (found in no more than 100 locales, or numbering no more than 10 000 individuals).

Whereas information on species used to be scattered around countless sources of published and unpublished journals, it is now all collected in one place. The data base has been described as a 'customised computer filing cabinet.' Information on any rare species in a given region can be retrieved from the data base for interested parties. In most cases, the person or organization seeking information is planning to develop an area and wishes to minimize the environmental impact of the development. Requests for such information are commonly made by the Army Corps of Engineers, the Department of Transportation, National Forests, county agencies and private citizens. The data base, by providing a form of 'one-stop shopping', may encourage prospective developers to seek information on an area's natural environment at the development's inception. Adjustments to the development plan, to allow for conservation measures, are more likely to be incorporated at an early stage rather than later on in the construction process. For ease of identification, some information has been recorded using a mapping and image processing system – a commercial, digital mapping program that transforms data into colour maps, showing the areas and boundaries of a particular species' habitat.

Apart from providing an invaluable source of information for other parties, the Natural Heritage Program enables the Nature Conservancy to make informed decisions concerning which areas of land to protect. The development of the Conservancy's selection criteria is attributed to Dr Robert Jenkins, who joined the Conservancy in 1970. At that time the Conservancy appears to have had no explicit criteria for selecting sites for protection: 'They weighed their decisions on prettiness, utility and recreation – huntin', fishin', boatin' and berry-pickin' . . . we needed some yardsticks, some standards, some consistent measurements.' (The Nature Conservancy, 1992).

The Conservancy had already carried out a great deal of inventory work by 1970 but had organized information by area, rather than species. Jenkins sought information on which areas housed the most imperilled species. By inverting the computer files the species themselves became the focus of the data, rather than each area. The data base then enabled the Conservancy to systematically rank areas according to their respective wealth of rare species and other ecological entities. The Heritage Network encourages and enables new information to be sought concerning species. Regularly, species thought to be extinct reappear. For example, the northern bog lemming (*Synapotmys*) has been seen recently in Maine for the first time in 90 years.

The Heritage Network has recently begun to tackle the difficult task of surveying habitats of migrating birds. In 1991, using volunteers and professional ornithologists, the Heritage Program inventoried North America's southbound songbirds as they funnelled through the mid-Atlantic coast. Four states were covered by the inventory – Virginia, New Jersey, Delaware and Maryland. It is hoped that the data will reveal critical areas of the landscape where the declining birds' habitats most need protecting.

Two areas for expansion of the Heritage Program are being looked at for the future. First, it is hoped that the Program will expand beyond the United States borders: a goal particularly important for the acquisition of better information on the habitat of migrating birds. Second, the Conservancy has identified the need to 'plug any holes' existing in terms of information gathered about the United States landscape. In particular, there appears to be information lacking on public lands, including federal parks, forests and rangelands. This will enable better information to be supplied to public agencies, hopefully to encourage improved stewardship of land. For example, about 35 million acres of land are held and managed by the Department of Defense in the USA. On some areas of this land The Nature Conservancy is actively pursuing the negotiation of management agreements for conservation of wildlife habitat.

It is hoped that the Heritage Network will be developed to include data on natural communities and stewardship-related issues, and that it will be extended to include new information and design tools that will address larger landscapes, regional and national perspectives. Once the inventory has identified an area of habitat that supports a rare or imperilled species, Conservancy staff seek some sort of protection for the site (see Fig. 3.3). The Conservancy has recently been extending its range of protection measures to supplement its traditional approach of the acquisition and retention of sites, negotiating a whole host of protection mechanisms based on a more diverse range of property rights than freehold ownership (see Chapter 4).

Fig. 3.3. The local, low-key approach: a Nature Conservancy staff member negotiates an easement with a landowner in Wisconsin. Photograph: Victoria Edwards.

Ensuring protection: the stewardship program

Since its inception, the Nature Conservancy has placed emphasis on the establishment of preserves. However, in the last few years it has recognized the importance of restoring habitat and monitoring and protecting species not only on its own land, but also on other private and public land. The steward-ship scheme pursues the Conservancy's aims with the specific emphasis on ensuring that the rare species and natural communities are preserved over time. To do so, stewardship aims to:

identify what the rare communities and species need to survive on each
  area of land;
provide those needs; and
monitor the health of those life forms to ensure success.

The stewardship scheme covers all Conservancy-owned land and all land covered by an agreement with the Conservancy (such as easements, leases and management agreements). As such, 1500 preserves are included covering over 1 million acres of land. In addition, a further 300 000 acres of public land, managed cooperatively with the Conservancy, are included in the stew-ardship scheme. Approximately 200 full-time staff are employed in steward-ship, with 'interns' employed for short-term work on specific tasks and a

Fig. 3.4. Prairie remnant, Minnesota: remnants often exist where rocks and boulders have prevented agricultural cultivation. Photograph: Victoria Edwards.

whole host of volunteers used for a variety of tasks. Often one-off, simple tasks will achieve the desired protection. For example, when a rare locoweed in northern California needed protection from browsing animals Conservancy volunteers and staff fenced off the plant, resulting in an increase from 88 to 15 000 seed pods in one year. In other instances, stewardship calls for a permanent, on-site manager to restore habitats and monitor progress. Brian Winter, the Director of Stewardship and Manager of the Conservancy's Western Preserves in Minnesota, is restoring native tallgrass prairies through the collection of seed and experimentation with cultivations. The Conservancy manages more than 17 400 acres on 65 preserves in Minnesota, including the greatest number of privately owned and managed prairie preserves in the USA (Fig. 3.4). In 1975, the Minnesota Chapter established a prescribed burn program: an average of 5000 acres of native prairie are burned annually on the Conservancy's Minnesota lands.

Case study: Spring Green Preserve, Wisconsin

Wisconsin has a strong association with the conservation of natural areas: it has been home to two of America's most famous naturalists, Aldo Leopold and John Muir. The Wisconsin Chapter of the Nature Conservancy was founded by a small group of scientists and conservationists in 1960. Working mostly with volunteer staff in its early years, the Wisconsin Chapter now has a full-time staff of 17 and by 1994 was responsible for protecting 128 reserves, covering some 48 204 acres of land in Wisconsin, and had a membership of almost 20 000 people.

The Spring Green Preserve in Sauk County, Wisconsin, covers some 481 acres and encompasses Wisconsin's highest quality and largest representation of sand prairies: a complex community of dry-lime prairie with oak openings. Two reptiles, the slender glass lizard (*Ophisaurus attenuatus attenuatus*) and the ornate box turtle (*Terrapene ornata ornata*) are also found here.

The 481 acre property comprises 260 acres purchased and owned by the Nature Conservancy and a further 221 acres leased by the Conservancy from private landowners for nature conservation management. In addition to the 'management lease' the Conservancy has obtained a 'right of first refusal' over the privately owned parcel of land. This particular acquisition tool requires that the landowners provide the Conservancy with the opportunity to purchase the land before it is offered for sale elsewhere. Land situated beyond the high quality natural community occurrences were included in the lease, to act as buffer zones in which to establish fire breaks for prescribed burns. Stewardship advice was supplied to the landowners before any formal property rights were negotiated.

The Spring Green Preserve, currently used for naturalist tours and some deer hunting, is typical of the vast collection of small preserves that the Conservancy has built up over its 40 years.

Land trusts

In the mid-1800s, many 'village improvement societies' were formed in New England to 'improve the quality of life and of the environment' (Land Trust Alliance, 1990). These small, nonprofit organizations were the forerunners of today's land trust movement. In 1891, the Massachusetts legislature incorporated 'The Trustees of Reservations' to protect the 'jewels of the living landscape.' It was this first land trust which was later to be used as the model for the 'National Trust for Places of Historic Interest or Natural Beauty in England' (which adopted its articles of establishment directly from the Trustees of Reservations in 1894).

Land trusts are local, regional or statewide nonprofit organizations, directly involved in protecting important areas of private land. Land trusts are not 'trusts' in a legal sense: indeed, many refer to themselves as 'conservancies', 'associations' or 'foundations'.

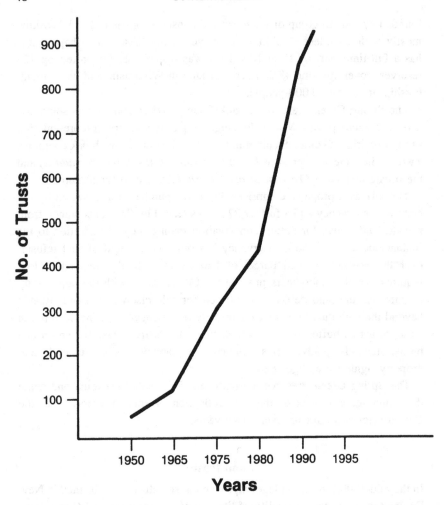

Fig. 3.5. Formation of land trusts. Source: Land Trust Alliance, Washington, DC, personal communication.

The USA has almost 1000 land trusts, which have protected some three million acres of land in total. They have been referred to as America's fastest growing conservation group: almost a half of the land trusts in existence have been formed in the last decade. Currently, new land trusts are being formed at the rate of more than one per week (Fig. 3.5). They vary considerably in size: over half are small trusts run solely by volunteers, while others have a large professional staff. Annual budgets range from under $10 000 to over $1 million (32% operate with budgets of $100 000 per annum or more).

Most land trusts derive their income from membership fees and donations. Collectively, they have nearly one million members. Other funds come from government agencies, corporations and foundations (there are 43 private foundations in the USA specifically interested in funding 'green space'). Like The Nature Conservancy, they have nonprofit, tax-exempt status. Land trusts also borrow money to protect land. Loans are repaid either through fund raising, or sales of land to conservation buyers or state and federal agencies when public funds are available. Each land trust has provision for another trust to take over its conservation responsibilities should it fall into financial difficulty.

Protection strategy

Whereas the Nature Conservancy has the single focus of protecting biological diversity, land trusts protect land with a wide variety of values – natural, recreational, scenic, historic or productive – depending on the needs of the local community. Like the Conservancy, the trusts use a variety of mechanisms to protect land. In total, land trusts own 437 000 acres and hold easements over a further 450 000 acres. Land trusts also acquire land to ensure its protection and then transfer it to another organization or government agency. Over 668 000 acres have been protected in this way.

While the Nature Conservancy relies on a national inventory system to collect information for the design of its protection strategy, land trusts rely on their decentralized operation: their presence in the local community enables them to seek and react to information regarding a property's natural qualities and its status of protection. Generally, therefore, the trusts tend to be more reactive than proactive, responding to threats to natural areas rather than targeting specific sites. As a result of their local presence, land trusts are able to offer quick, flexible responses to threats to private land. They tend to reflect the values of the immediate community, rather than hold a national perspective. In this respect, they offer a complementary service to an organization such as the Nature Conservancy, which seeks to protect global and nationally significant natural communities.

Case study: the Trust for Appalachian Trail Lands

The Appalachian National Scenic Trail is a public footpath across 2155 miles of the Appalachian Mountain ridgelines, stretching through 14 states, from Maine in the north to Georgia in the south (Figs. 3.6 and 3.7). The Trail evolved from the proposals of a Massachusetts regional planner, Benton

Fig. 3.6. View from the Appalachian Trail, West Virginia. Photograph: Victoria Edwards.

Fig. 3.7. Walkers on the Appalachian Trail, West Virginia. Photograph: Victoria Edwards.

Mackaye, who wanted to preserve the Appalachian ridges as an accessible natural belt that would serve as a retreat for eastern urban dwellers. It was designed, constructed and marked in the 1920s and 1930s by volunteer hiking clubs and the Depression-era Civilian Conservation Corps, coordinated by the Appalachian Trail Conference (ATC), which was formed in 1925 to coordinate volunteer effort and to monitor, maintain and protect the Trail and the land it crosses.

Originally, much of the Trail passed over private land with the consent of the owners: 'In the early days, such agreements required nothing more than a handshake, a sense of goodwill, and a promise to close the gates.' (Appalachian Trail Conference, 1990). However, when the 1960s brought development along the Trail and a change in character from rural to suburban in many regions, keeping the trail open as a continuous route meant relocating the route from one piece of private land to another. By 1968, at least 200 miles of the Trail had been relocated to roads and it was feared that the Trail would soon be lost.

In 1968, Congress recognized the problem and passed the National Trails System Act (PL90–546). The Act designated the Appalachian Trail as the nation's first 'National Scenic Trail' (there are now eight in the USA) and authorized federal acquisition of land, or interests in land, along the route.

Essentially, the Trail became a linear National Park administered by the National Park Service (US Department of the Interior), the Forest Service (US Department of Agriculture) and state governments. The aim was to provide a permanent right-of-way, limited to approximately 100 feet either side of the Trail. Land was acquired by both federal agencies and the governments of the states through which the Trail passed. The Appalachian Trail Conference, through direct work and through coordination of other local nonprofit organizations, has assisted in the purchase of land for the Trail's main corridor. Land is added to the trail by negotiating with private landowners to acquire lands or rights to land and transferring them to government agencies. In 1982, the ATC established the Trust for Appalachian Trail Lands, a private land trust which works like any other land trust, but with a specific focus on the Trail.

The Trust uses private means and donations to acquire land or easements over land. The Trust's first priority was to help the National Park Service and other agencies acquire land for the Trail's corridor. The Trust's role in the acquisition program was the result of recognizing that the flexibility of a private organization may facilitate difficult transactions. By the end of 1993, the Trust had been instrumental in protecting over 13 000 acres of land. Within a few years, the land acquisition phase for federal and state agencies will be over and the Trust's focus will broaden to lands around the Trail.

Amendments to the Trails Systems Act in 1978 and 1983 increased the maximum design width of the corridor to an average of 500 feet on either side of the Trail. However, there remained concern that development immediately outside the corridor would impinge on the recreational experience of walkers on the Trail. Indeed, at some sites the Trail is only 50 feet wide and where the National Park Service has delegated authority to the states to acquire land for the corridor, the right of way may be as narrow as 40 feet. These narrow rights of way do not allow the flexibility managers require for any necessary realignments of the path and for control of erosion and drainage. There was a fear that the Trail would become 'a thin green ribbon winding through an increasingly developed landscape.' (Appalachian Trail Conference, 1990). The Trust for Appalachian Trail Lands, therefore, began to focus its attention on 'supplemental lands' on either side of the trail. These lands, which were not included in the Trail's main corridor, could contribute to the 'visual, functional or ecological value of the Trail experience.'

The Trust works directly through the purchase of land and easements, and indirectly by stimulating local land trusts and affiliated clubs by providing them with grants and technical assistance. In addition, recommendations for the Trust's own action or specific projects are often solicited from local Trail club members and local land trusts. In 1983, the Trust published a set of guidelines for local groups, identifying the types of property it may wish to protect. These range from land that is important for protection of wildlife habitat, to land that has development or mineral extraction potential (which might detract from the enjoyment of the Trail) and land that could provide access to an important drinking water source, shelter or campsite.

Case study: Gallatin Valley Land Trust

Gallatin Valley Land Trust, being established in July 1990, is among one of the more recently formed land trusts in the USA. It is a private, nonprofit, tax-exempt organization and is based in Bozeman, Montana. The Trust was formed specifically 'for the conservation of scenic open space, agricultural land, historic sites, recreational opportunities and wild-life habitat.' The Trust is directed by a 15-member board of directors and an additional 10-member advisory board, supplying specialist skills such as accountancy, architecture and ecology. Both boards comprise local, voluntary members. The Trust is managed by the full-time Executive Director, Chris Boyd, with part-time support staff. For capital resources the Trust relies on local contributions, land transaction fees and a limited number of foundation and government grants. Most of the land protected is achieved through the donation of conservation easements, although the Trust may also receive gifts of land. In addition, the Trust provides information and advice on land conservation techniques to landowners, property buyers and public agencies.

Early in its development, the Trust identified the need to clarify its own conservation priorities. It has since experimented with an evaluation checkl-ist, containing 11 criteria and calling for a score for each aspect of a piece of land. The Trust focuses its efforts on local, citizen-initiated projects and particularly on projects located along the edges of communities. The Trust keeps in close contact with other conservation groups in its geographic area, such as the Greater Yellowstone Coalition, Montana Land Reliance, The Nature Conservancy and the Sacajawea Audubon Club.

Since its inception in 1990, Gallatin Valley Land Trust has entered into six conservation easements totalling 1200 acres of land. The Trust's projects, which center around the town of Bozeman, include: two public access ease-ments along a creek; a recreation easement with cross country ski trails in the Bridger Mountains; a neighborhood trail system known as 'Main Street to the Mountains'; a 'rails-to-trails' project along the Missouri River, converting redundant railway line to footpaths; and the establishment of a public park close to Bozeman encompassing a traditional sledding hill and a prominent ridgeline overlooking the valley.

In addition, conservation easements have been entered into on agricultural properties. One hundred and eighty acres of private land have been protected from subdivision and development at the wishes of the present owner, 87-year-old Gertrude Baker. Baker, who was concerned about what would happen to the land after she died, wanted to protect it as wildlife habitat for visiting elk, moose, deer, squirrels and other animals and birds. Apart from the usual restrictions of development prohibitions, a number of restrictions have been incorporated in the easement, specifically to protect the wildlife in accordance with the owner's wishes. For example, any future owners of the property will not be permitted to own a cat, because of the danger of cats catching and killing chipmunks. In drawing up the easement the Trust had an unenviable task of trying to predict any future behavior that might pose a threat to the present owner's wishes. Baker did not want the land to be locked away for wildlife, but was keen to ensure that it would still provide a home for any future owner: as Chris Boyd stated, 'She has the feeling that with people on the property, there won't be vandalism and the cows that get in can be chased off.' (*Bozeman Daily Chronicle*, 1991). Several experts were enlisted to help and the Trust now believes that it has an easement that is flexible but represents Baker's wishes for the land.

The Trust is being approached by increasing numbers of landowners seeking information about conservation easements on their land. Inquiries seem to be predominantly from local citizens and landowners, whose interest in easements stems from a desire to protect the area from subdivision and housing development: 'our community's awareness of development pressures in the Gallatin Valley is growing'. (Gallatin Valley Land Trust, 1994).

Case study: Montana Land Reliance

The Montana Land Reliance (MLR), a statewide land trust, was founded in 1978 and, to date, has protected over 123 000 acres of land with 91 conservation easements. It is a charitable land trust, which relies on private contributions for its financial support. Gifts are tax deductible, since the MLR is a nonprofit, tax-exempt organization. Forty-four of the easements and 52 600 acres of the land are situated in the Greater Yellowstone Ecosystem, an area of land surrounding Yellowstone National Park that has been identified as necessary for the Park's successful protection.

The mission of the MLR is 'to permanently protect private lands that are ecologically significant for agricultural production, fish and wildlife habitat and open space.' Local landowners are encouraged to donate conservation easements to the Reliance. Landowners who may wish to protect special features on their land, but are not ready to donate property rights to the MLR, may join its 'Conservation Partners'. Essentially, these landowners recognize the need to demonstrate good stewardship of natural areas but do not wish to enter into a formal agreement. It is often seen as the first step in accepting the idea of donating a conservation easement.

Rather than protect biological diversity, the MLR aims to protect Montana's farmland and open spaces, its fisheries habitat and its wildlife habitat on a watershed basis. The work of the MLR is both reactive, responding to offers from private landowners to donate conservation easements, and proactive, mapping targeted watersheds and introducing conservation ideas to landowners.

The MLR recognizes that having more than one land trust in a state can be confusing to landowners and can create competition for limited funding sources. Nevertheless, with good communication and clearly defined missions, different land trusts can act in a complementary fashion, ensuring that the conservation requirements of the state are covered. Many economists theorize that it is this element of competition which ensures that private conservation organizations work effectively and efficiently towards their goals.

Conclusion: lowering transaction costs

A fully functional market system ensures that a diverse range of products can be produced and traded to match the diverse demands of different groups. In conservation this can be achieved through the development of an increasingly elaborate system of property rights that allow natural areas to be protected, and through the production of goods and services that satisfy the demands of those who are willing to pay for protection. Development of the range of protection mechanisms and the range of products can be expensive. Private conservation organizations can, however, reduce the overall costs of delivery by facilitating collective action. This chapter has explained how The Nature Conservancy has taken advantage of the economies of scale available to a national conservation organization, while ensuring that it is still represented at the state and local level. In particular, it has developed a comprehensive resource inventory program that provides information for private and public agencies throughout the USA.

As small independent organizations, the land trusts face the problem of transaction costs in their attempts to protect private land. First, they need to have extensive knowledge of the laws governing conservation protection mechanisms, other property transactions and federal and state tax laws. Second, they do not benefit from economies of scale, which an organization such as The Nature Conservancy enjoys. For example, the costs of research into conservation and land protection and the costs of providing services, such as liability and property insurance for protected areas, have the potential to be very high. Third, whereas a large organization such as the Conservancy might have enough capital funds to ensure immediate protection of an area about to be sold or developed, land trusts, often relying on a small resource base, may not be able to act immediately. This is an important constraint, since time is often of the essence in land protection.

Rather than give in to the high transaction costs of being small independent organizations, and leaving protection of land to large public or private sector bodies, the trusts developed two solutions. The first was to unify under an 'umbrella' organization and yet retain their independence. The second was to use a national organization with similar objectives to help with fund raising.

In 1982, the Land Trust Alliance (LTA) was founded to provide leadership and services to the trusts. It provides specialized services, publications, information and training for land trusts and other land conservation organizations. Through its 'Umbrella Insurance Plan' it has made liability insurance more affordable for nearly 200 land trusts. It has also prepared outline policies and procedures to guide land trusts and help them to ensure that their operation and land transactions are 'fundamentally sound'. This 'Standards and Practices' document covers 15 key aspects of the land trusts' operations from, for example, the establishment of goals and purposes of land trusts, through to board accountability, conservation methods and stewardship (Land Trust Alliance, 1989).

In addition, land trusts may take advantage of The Trust for Public Land, which was founded in 1972 as a national conservation group specifically to conserve land for public recreation and open space. The Trust for Public Land is not an advocacy or membership organization: over half of its support comes from its land transactions, the remainder of its revenue comes from individual gifts, foundations and corporations. The Trust for Public Land works with government agencies, land trusts and other conservation groups to protect land by bridging the resource needs of these groups. When land is placed on the open market and funding cannot be found by a local group, the Trust negotiates a real estate deal to quickly acquire and protect the land from development until funds are found. Essentially, it acts as a lending bank to the smaller land trusts and government agencies. It has a large revolving fund, which enables it to act swiftly to purchase land until funds can be raised for permanent protection. The Trust also offers training and technical assistance to land trusts with regards to real estate negotiation, joint venture land acquisitions and tax deductions.

Land trusts, by combining under an umbrella organization, have strengthened their ability to reduce the transaction costs of protecting private land while retaining their local profiles. Together, land trusts have improved the availability of funding and designed insurance provisions for operation. All conservation organizations are continually seeking new mechanisms by which to protect natural areas and incentives to augment the use of such protection. Chapter 4 examines the type of mechanisms available to the conservation organizations and analyzes the effectiveness of the current incentive structure for land protection.

4

Protection mechanisms and incentives

> Every species, every habitat, every ecosystem presents [the
> Conservancy with] a different challenge. Our conservation
> strategy must vary accordingly.... We must, in short, be able
> to match our protection tools with the task at hand.
>
> *The Nature Conservancy, 1989*

Protection mechanisms

The success of conservation on private land in the United States has often
been attributed to the ability of conservation organizations to act in a flexible
manner. Conservation of natural habitat on private land in the USA relies on
a number of formal and informal arrangements to ensure proper protection
and management of the land:

fee simple ownership (through purchase, bargain sale or donation);
conservation easement (through purchase or donation);
management agreement;
lease; and
informal agreement.

The suitability of each type of mechanism depends upon the type of habitat
being protected and the type and distribution of values associated with the
site. In most cases some form of property right is acquired from the landowner's interest, in order to achieve protection of a natural area. The interest
held by a landowner is referred to as a 'fee simple' interest. The imagery of
a 'bundle of sticks' is often used to explain the interest held, where each of
the sticks represents a right associated with the property. The entire bundle
of sticks can be acquired by acquiring the fee simple interest. Fee simple
ownership affords the ultimate protection to an individual or organization
wishing to conserve natural habitat. Alternatively, where other compatible
uses might be made of the land and there is no wish to hold the land in
absolute ownership, certain rights can be separated from the 'dominant
estate', like sticks from a bundle, and transferred to the conservation organization as 'less than fee' interests. An easement is one such 'less than fee'
interest. Easements are generally restrictive in nature. They may be supplemented by a management agreement that can afford the holding organiz-

ation greater input into the day-to-day management of the property. In some cases a formal contract or lease may be used where protection of the wildlife habitat depends upon positive conservation management practices. Finally, where it is perceived that the costs of formally protecting land are not necessarily justified, a simple, informal agreement might be made between the landowner and an interested second party. This may take the form of a signed letter, written agreement or merely a handshake and does not constitute any transference of property rights.

Fee simple ownership

Land might be acquired by a conservation organization by gift, bequest or purchase (at full or less than market value). While normally land is transferred with vacant possession, an individual landowner may gift land to a conservation organization but reserve the right to retain possession of the property for his/her lifetime or for the lifetime of immediate family. In effect, the donor is gifting the reversionary interest in the land after retaining a life interest.

Ownership of land is often cited as the ultimate protection mechanism, since it affords the conservation organization the most security of protection while allowing flexibility in management practices. It is, however, an expensive form of protection: even if land has been donated, bequeathed or sold below market value to the conservation organization, owning land and managing it in perpetuity requires a substantial amount of resources. For this reason, organizations such as the Nature Conservancy often insist that endowments accompany donations of land, to cover future management expenses. In addition, an organization should be aware of the opportunity cost of holding land that might be protected by some other means or some other organization. Placing a conservation easement over the land and selling the property on to a third party releases some of the capital tied up in the property. This option, however, often results in less management control for the conservation organization. Where the future management of an area of rare species or natural communities is critical to its protection, then an organization may be especially reluctant to part with the freehold title. Reluctance may be based not only on the observation that no other organization can ensure protection of the natural area, but also that the landowner can learn from the experience of ownership. David Carr, manager of The Nature Conservancy's Pine Butte Preserve in Montana, believes that one of the main advantages of retaining and managing the preserve is that the Conservancy is still learning about its sensitive ecosystems and how best they can be protected. The

Conservancy has been criticized for its practice of selling reserves to government agencies for future management (see Chapter 9).

Often conservation organizations will hold land in fee simple ownership in order to meet some policy objectives other than the protection of that specific piece of land. For example, a conservation organization may hold land as:

a showcase: where ownership and management of the land is important to create a good example to other land managers in the area; or

an operational asset: where land is close to an urban area and presents a good site for entertaining and demonstrating the organization's work to potential donors, or for the operation of field trips for members of the local community.

Easements

An easement is a contract between a landowner and some second party by which restrictions are placed over the future use of the land, thus protecting the land's qualities. The effect is for the landowner to vest an interest in the land in the second party. An easement is exclusive and unique to the land to which it is attached. The owner continues to manage the land in accordance with the conditions of the easement, and the second party becomes the overseer to enforce the easement for its duration.

Conservation easements were established by statute in the USA and, as such, differ from easements ('covenants') at common law in several respects. Easements may exist in perpetuity or for a specified period of time. Unless a conservation easement is granted to a qualified conservation organization in perpetuity, however, no charitable deduction will be allowed for federal income tax purposes (see p. 64). As a consequence most easements are perpetual and therefore afford permanent protection. Unlike restrictive covenants at English common law, they do not require the presence of another piece of adjacent land to benefit from the easement and they can pass to successors in title to the land. The landowner who enters into the easement can use the relevant land, subject to the restrictions imposed, and can sell or otherwise dispose of the land. The landowner is still liable to pay property taxes and to ensure that the restrictions are not violated. However, the easement is registered on the land's title and when the land changes ownership it remains subject to the conditions of the easement. The rights, which the landowner relinquishes, are transferred to an organization or body, such as a conservation organization or government agency, by virtue of a legal document. In addition to the right to enforce the restrictions placed on the land, the easement holder generally has limited rights of access for inspection, scientific data collection or other purposes agreed to by the landowner. Easements are transferable and can be sold or donated from one organization to another.

Generally, the conservation easement is a restrictive document. If the land requires active management to preserve or restore natural habitat, some management rights may be granted to the easement holder. Each easement and each set of restrictions is specific to the protection needs of a particular site. In some cases the easement may provide that the land is left completely in its natural state. In others the easement may allow activities such as hunting, grazing or even limited development (see Chapter 7), provided that it does not destroy the ecological value of the land.

The easement can apply to the entire property or to part of it. An inventory is normally conducted upon signing of the easement, comprising detailed documentation, maps and photographs of the land's condition. This is then used as a benchmark from which to measure the success of future protection and management of the site. The holding conservation organization is responsible for monitoring and enforcing the provisions of the easement. This is done by routine visits to the site. If the terms of the easement are breached, restoration of the property to its prior condition is sought. Generally, easements will be enforced by seeking a court injunction and/or damages if irreversible damage has occurred. The costs of such a procedure are often high. In order to avoid such costs, some easements prescribe binding arbitration if a disagreement arises or prescribe how costs will be allocated if litigation is necessary. Others insist upon an endowment to accompany the easement, which will help the organization defray some of the costs of monitoring and enforcing the easement.

There are several problems associated with the use of easements as a protection mechanism. First, it is possible that violation of easements will increase with time, particularly if the land passes to new owners. However, some land trusts have commented that since new owners are aware of the presence of easements (and may in fact acquire the land because of the protection afforded to it by the easement) they may be just as sympathetic to its objectives as the original owner. In its 1985 study of easements, the Land Trust Alliance found that while 13% of government programs reported violation of easements, only 5% of private programs had expressed any problems of violation (60 cases in total). The report also stated that 'whether or not the easement was donated or purchased does not appear significant in regard to the likelihood of violation.' (Land Trust Alliance, 1985).

Second, an easement does not protect that land from 'condemnation'. In the USA, the government has the right to take land or property rights in private land through the right of 'eminent domain'. Eminent domain is the right of the government to take private property with just compensation for public use through 'condemnation' proceedings (known as 'compulsory purchase and compensation proceedings' in Britain). Federal, state and local governments, as well as special purpose government agencies such as school boards, public service districts, soil conservation districts and utility companies, exercise powers of eminent domain. However, there are certain exceptions to this rule. First, gift of an easement to federal government will protect the land from a state's power of eminent domain. Second, some state laws (such as the Natural Resources Article 51202 of Maryland) insulate land with a conservation easement from condemnation for other limited, open space purposes. Third, some state laws will not allow a county or other subdivision to condemn a conservation easement (Chesapeake Bay Foundation & University of Maryland School of Law, Law Clinic, 1988).

Leases

A landowner may grant a lease to a conservation organization, which will allow the lessee the right to enter and manage the land according to agreed terms. A lease may be an appropriate mechanism when a landowner demands monetary compensation for restricted land use and where financial resources allow a long-term commitment to payments. In particular, a lease may be used as a short-term mechanism, until a more permanent solution may be found. The proactive nature of the lease agreement would also render it especially appropriate where restoration work may need to be carried out in a natural area. Since there are few monetary incentives to be gained from donating a lease, donating a conservation easement would normally be seen as more preferable to the landowner. The permanent protection afforded by easements generally makes them more highly sought by conservation organizations than leases.

Management agreements

A management agreement is a less formal way of enlisting the help of a conservation organization in protecting natural habitat on private land. A conservation plan is drawn up for the property and the owners agree to a specific course of action. The conservation organization provides technical expertise and assumes some of the capital and labor costs involved. This mechanism is also appropriate for meeting habitat restoration objectives. Management agreements can usually be cancelled by either party at 30 days notice. While not creating permanent protection of a natural area, management agreements are a useful tool for introducing landowners to the concepts of protection and to suitable management practices.

Gentle persuasion: natural area registers

The Nature Conservancy has a program whereby landowners agree to voluntarily protect rare plant and animal species occurring on their properties. Some state chapters hold a 'Registry of Natural Areas' to 'recognize, honor and aid' landowners who have protected special areas of biological diversity or natural communities on their properties. By registering their land, landowners make a verbal agreement to

> continue the practices that have preserved the special features of the land up until now;
> notify the Conservancy of any possible threat to the area such as pollution, rights of way, drainage etc.;
> notify the Conservancy of any intention to sell or transfer ownership of the property.
>
> *The Nature Conservancy, 1989a*

Since the registry agreement is voluntary and oral, it might be cancelled at any time by a landowner. There are no government regulations attached, no legally binding agreement, no access rights are granted and directions to the site are not published. There is no charge, nor payment for a landowner enrolling in the registry scheme. The Conservancy will provide management advice for the area, free of charge, and present each owner with a wooden plaque, inscribed with the landowner's name and the name of the natural area. The Conservancy uses the registry in one of two ways. First, as a means of persuading landowners to enter into a conservation easement, management agreement, bargain sale or gift of land. Approaching the landowner with an oral, nonbinding agreement is often a very effective means of obtaining more secure protection at a later date, when the Conservancy has earned the landowner's trust and respect. Second, it enables the Conservancy to encourage sound management practices on land that may not warrant a full management agreement or conservation easement (the area may be too small or of insufficient conservation value). From the landowners' perspective, the registry program often initiates a gradual process of realization and understanding of the value of wildlife habitat on their land. The Wisconsin Chapter of the Nature Conservancy cites one landowner: 'The registration made us more aware that what we have was unique. Before that the land had just sat down there for years. Now we realize we have something really worth saving.' (The Nature Conservancy, 1989a). A 'Natural Heritage Stewardship

Program' operated in Ontario, Canada, found that approximately 20% of landowners in Southern Ontario needed no further incentive to protect their natural area than the provision of information about the area and its environmental quality (Hilts & Moull, 1986). Many landowners are quick to point out that restricting future management of the land is a seemingly ironic response on the part of the conservation organization if the species have survived past and present management practice. The registry program, therefore, works on the basis of rewarding good stewardship and monitoring protection of the species.

The incentive structure

Historically, conservation organizations in the United States have relied upon the gift of land and conservation easements as a means of protecting private land. More than half of the Nature Conservancy's preserves were donated to the organization. Apparent generosity of this type begs the question 'what do the landowners benefit from donating a conservation easement?' First, when making a gift of land or a gift of an easement, the landowner is assured that the land will be protected in perpetuity in its natural state. Knowledge that the land will be protected beyond the landowner's lifetime provides private benefits to an individual landowner, which should not be underestimated. Research suggests that it is often a major factor in a landowner's decision to gift land (Edwards & Sharp, 1990). A survey conducted by the Land Trust Alliance in 1985 indicated that 'protecting the land they love' was the primary motivation for 67% of landowners donating conservation easements. Indeed, the Wisconsin Chapter of the Nature Conservancy obtains several enquiries each week from people, often with no children, who wish to ensure that their land will be protected in perpetuity (Braker, personal communication 1991).

Second, there are considerable tax advantages to be gained from gifting land or easements to a conservation organization. If land is gifted to a conservation organization, the value of the gift for tax purposes is its full and fair market value. The value of an easement is the difference between the value of the land without the easement and the value of the land after the easement has been granted, as determined by the nature of the restrictions imposed and their impact on present and future land use. The valuation must be carried out by a qualified appraiser and accepted by the Internal Revenue Service. Each parcel of land is unique and no definite average percentage of value attributed to the easement can be relied upon. The availability of tax deductions depends upon current state and federal laws. Generally, the following may be available:

1. Federal and State Income Tax Deductions. The provisions of the Internal Revenue Code of the United States enable individuals and corporations to deduct the full fair market value of gifts of land to 'qualified' conservation organizations on their federal income tax returns: 'qualified' organizations are defined by the Internal Revenue Code and regulations. Generally, a taxpayer is unlikely to be allowed a charitable deduction where less than the taxpayer's entire interest in the property is contributed. However, the Internal Revenue Code makes special provision in such cases for gifts of reversionary interests in land and conservation easements on conservation property. Usually, gifts of land and easements also qualify under state income tax laws as charitable contributions.

   There is a limitation that the value of the gift can only be deducted against up to 30% of the donor's adjusted gross income for the year of the donation (50% if the property is gifted in the same year as its acquisition). In both cases, the balance of the deduction may be carried over for five succeeding years, subject to a 30% of income limitation in each of those years.

   For example Assume that individual 'X' has a gross annual income of $120 000 p.a. and pays federal income tax at a marginal rate of 36%. X donates a property to a conservation organization. The property has a full market value of $240 000 and was originally purchased for $60 000. X can deduct $40 000 (30% of her gross income) each year for six years from her federal taxable income liability, until the value of the land donated ($240 000) has been fully deducted. At 36%, this represents a tax saving of $86 400.

   Similar benefits can be obtained by donating a conservation easement over the land.

   For example Y donates a conservation easement over some agricultural land to a local land trust. The easement restricts the amount of development that may take place on the property. An appraisal confirms that the property has a value of $600 000 before the easement is placed on the title. Restrictions imposed by the easement reduce the value of the property to $400 000. The potential deduction for income tax purposes is $200 000. Assuming that Y's gross annual income is $130 000 and that he has made no other charitable contributions in the year, his charitable deduction resulting from the donation is $39 000 (30% of his \$130 000 annual income). By carrying the unused portion of the donation forward for five years, Y will receive a total of \$200 000 in charitable deductions. At a tax rate of 36% this represents a total saving of \$72 000 in federal taxes. If Y's gross income increases over the six-year period, then he will be able to deduct greater amounts each year, as the 30% of gross income rule results in an increase in his allowable charitable deduction.

In the event that the property owner is a corporation, the gift of the easement may be deducted against 10% of the corporation's annual income in the year of donation, with any unused balance limited to 10% of the annual income for the next five years.

Whether donors will receive similar tax benefits from state income tax when donating land and conservation easements depends on the tax laws of each state. Some states have passed laws which eliminate all itemized deductions and replace them with a standard deduction. Under such laws, there would be no deduction for donations of conservation easements.

2. Federal Gift and Estate Taxes. In 1986, tax reform legislation separated the gift of a conservation easement from the gift and estate tax provisions. Conservation easements will normally result in a reduction of the property's value for estate and gift tax purposes. This can substantially reduce the financial burden of passing the property on to the landowners' heirs. In some cases, where the gift of a conservation easement fails to qualify as a federal income tax deduction (as above), it may qualify as a gift tax deduction. Environmental groups have been campaigning to have legislation reformed to provide full exemption from federal estate taxes for lands with publicly important conservation values on which the owners have donated a qualifying conservation easement (for example, the proposed Open Space Preservation Act 1993, which is still viable). Until such reforms, current estate tax provisions include an exemption equivalent of the first \$600 000 of an individual's assets. This, coupled with the ability to reduce the value of the land held under conservation easement, can enable some landowners to effectively eliminate the threat of estate taxes.

3. Federal Capital Gains Tax. Capital Gains Tax may be avoided when land is donated to an organization.

   For example In the previous example individual X donates a property to a conservation organization. The property was acquired for \$60 000 but now has a full market value of \$240 000. If instead of donating the property X had sold the property and donated the sale proceeds, the tax benefits would be reduced considerably. First, there would be a Federal Capital

Gains tax liability on $180 000 ($240 000–$60 000), which at 28% represents a payment of $50 400 and leaves only $189 600 of the sale proceeds for the conservation organization. In addition, X will only enjoy an income tax saving of £68 256 ($189 600 × 36%), as opposed to the $86 400 income tax saving received if she donates the land.

4. State Inheritance and Gift Taxes. Tax benefits available for state inheritance and gift taxes depend upon the type of taxes present in each state and the deductions allowable. Some states have no gift tax.

5. Property Taxes. State law, local practice and local tax assessors determine whether a conservation easement carries a reduction in the assessed value of the property. If the value of the property is reduced, real property taxes payable will be lowered. The immediate benefits to the landowner will depend upon the nature of the easement granted and the respective reduction in the value of the land. (For further reading of the operation of conservation easements and their tax benefits, see Montana Land Reliance & Land Trust Exchange, 1982; Land Trust Exchange, 1985; Diehl & Barrett, 1988; Small, 1988; Maryland Environmental Trust, 1989; Land Trust Alliance & National Trust for Historic Preservation, 1990.)

Case study: The Flying D Ranch, Montana

The Flying D Ranch is located along the Gallatin and Madison Rivers, near Bozeman, Montana (Fig. 4.1). It is a 130 000 acre ranch, of which 107 000 acres is in private ownership and the remaining 23 000 acres leased from government agencies. In 1989 the owner, Turner Enterprises Inc., donated a conservation easement to the Nature Conservancy over the 107 000 acres. It is thought to be the single largest conservation easement in the USA. While providing permanent protection to critical wildlife habitat, the easement is also structured to ensure the economic viability of the ranch operation. The easement is both restrictive and proactive in nature. It limits subdivision of the ranch to a maximum of four partitions and prohibits mining and commercial timber production. However, the easement allows grazing, farming and commercial hunting on the ranch.

A baseline inventory was taken on the establishment of the easement. Fifty range sites were cataloged for plants, productivity, climate, soil types, etc. The report will prove an invaluable benchmark against which the Conservancy can monitor and review the effects of management practices on the ranch.

Some of the ranch's farmland is leased, from year to year, to four separate tenants for the production of crops. The tenants are all bound by the

Fig. 4.1. The Flying D Ranch, Montana. Photograph: Bud Griffith.

conservation easement. This means, for example, that before applying any pesticides to the land, the tenants must first gain the Conservancy's approval of type, dosage and means of application. The rest of the land is currently grazed by the ranch's 3300 bison (Fig. 4.2). The grazing capacity of the ranch is monitored to determine an acceptable stocking rate for the bison. An area's stocking rate is expressed in terms of 'animal equivalent units' (AEU). An AEU is the amount of forage that is required to feed a 1000 lb animal over a given period of time. A census of grazing wildlife is conducted annually and numbers of livestock grazed are limited according to the total grazing population of the ranch. Measuring the ranch's grazing capacity enables the manager to vary the size of the bison herd according to the grazing requirements of other wild animals – particularly the elk. In this respect, livestock will not take grazing preference over wild animals. The bison, which produce good quality lean meat with low cholesterol values, are very much a commercial herd: heifers are sold for breeding and bulls sold after weaning, as yearlings or at 18 months.

The owner, who spends a fair amount of time at the ranch, was keen to see the bison grazing in as natural a state as possible. Many of the internal fences on the ranch were removed, allowing free movement of the bison around the ranch. This also complements the 'preservation of open space' goal of the conservation easement. At first, keeping bison on the ranch provided a potential problem for wildlife. The bison need to be contained on the ranch for health and economic reasons. However, any boundary fences must

Fig. 4.2. Bison at the Flying D Ranch, Montana. Photograph: Bud Griffith.

allow free movement of migrating wildlife. Since the usual barbed wire fence, used for cattle, is not sufficient to hold bison, an electric fence was devised by the ranch managers and the Conservancy's wildlife biologists. The 4-wire fence is 4 feet tall, with a top and middle live wire and second and bottom cold wire. Since the bottom wire is 18 inches off the ground, the fence will enable young deer and elk to pass under it and mature animals to pass over it. Like most new activities on the ranch, erecting the new fence had to be referred to the Conservancy for approval.

The easement allows hunting on the ranch, which was established as a fee-earning enterprise by the previous owner. Elk hunts are sold as $4\frac{1}{2}$ day guided hunts, with 30 hunts sold annually. Five hunters are accommodated at any one time and each is supplied with a guide. The price is $8500 per person. In addition, 200 hunters are allowed access to the ranch annually to cull cow elk populations. It is estimated that 2500–3000 elk might inhabit the ranch (the elk will migrate for 75–100 miles). The ranch also has good trout fishing, access to which is reserved by the owner.

The Nature Conservancy was delighted to accept the easement. Shortly after the donation, Bernie Hall of the Montana Chapter of the Nature Conservancy identified the Conservancy's interest: 'We were interested in it from the standpoint that we could keep a large piece of property whole with agricultural use and open space, so it doesn't get subject to subdivision pressure.' (*Bozeman Daily Chronicle*, 1991). This is particularly relevant to this property, which lies on the outskirts of a growing community. The value of the conservation easement to the owner may be considerable. First and foremost, conservation easements allow an owner to ensure protection of a site in perpetuity. While an individual is capable of pursuing conservation objectives through sound land management practice during his or her lifetime, it is impossible to protect the land in perpetuity without the assistance of a perpetual body, such as the Nature Conservancy. The value to landowners of knowing that the benefits, which they associate with protected land, will be preserved for future generations should not be underestimated.

Conclusion: improving the market

By using common mechanisms for protection, the Nature Conservancy and local land trusts have been very successful in reducing the transaction costs of negotiating protection of private land. Both the Nature Conservancy and land trusts continue to portray themselves as nonpolitical groups, who use their funds to directly acquire property rights to land, rather than lobby government for greater land protection. However, both continue (albeit in a low-key manner) to lobby government for any changes in legislation, which they perceive will improve their ability to carry out their work. Primarily, effort is directed towards improving the overall efficiency of the conservation market by demanding changes in its institutional arrangement. This may involve lobbying for changes in the laws governing particular protection mechanisms or advocating greater public support, in the form of tax subsidies and grants to augment protection.

For example, for a while the statutory and common law rules of conservation easements differed in each state in the USA. In some states, a conservation easement was not perpetually enforceable unless the recipient owned adjacent lands that benefited from the conservation easement. Conservation organizations encouraged states to pass legislation which eliminated this problem. In 1981 the National Conference of Commissioners on Uniform State Laws drafted a Uniform Conservation Easement Act at its annual conference. The Act was approved by the Conference and recommended for enactment in all states. It was also approved by the American Bar Association. The Act has subsequently been enacted in several states. For example, legislation, which was passed in Wyoming in the early 1980s, enables qualified conservation organizations to hold easements in perpetuity. Prior to that date, the organizations would arrange for one acre of land to be transferred to them in fee simple along with the easement over the remainder of the land. Agreement was made with landowners to transfer the single acre back to the owner when the legislation was passed. By virtue of the lack of similar legislation, conservation organizations in a very few states are still forced to enter into these clumsy arrangements but the campaign is still in force to have such laws revised.

A second institutional improvement might involve reviewing the rules of condemnation. As stated earlier, an easement does not protect land from condemnation. While gift of an easement to federal government will protect the land from a state's power of eminent domain, gift of a similar easement to a conservation organization will not necessarily afford the same protection. Review of condemnation legislation might ensure that conservation organizations benefit from the same powers of protection as federal government agencies.

5

Fee-hunting

Recreational development is a job not of building roads into
lovely country, but of building receptivity into the still unlovely
human mind.

Aldo Leopold (Conservation Esthetic)

Introduction: conservation and recreation

In the late nineteenth century, about 44% of the population of the USA lived
and worked on farms. By 1950 numbers had decreased to 15% and in 1985
farmers represented a mere 2.2% of the population (USDA, 1986). While the
number of people farming may be decreasing, the total amount of agricultural
produce is increasing. Estimates suggest that by the year 2000, domestic and
foreign demand for food and fibre products could be met by nearly half
(218 million acres) the current cropland. With less reliance on rural land for
agricultural production, landowners are looking for alternative economic uses
for their land. The increasing urban population may hold the secret of the
success of alternative enterprises where improvement of land for wildlife
habitat is compatible with recreational use.

Increases in personal disposable income, vacation entitlements and life
expectancy beyond retirement, coupled with more comfortable, faster and
less expensive means of transport suggests that recreational demand will continue to increase worldwide. The benefits of recreation for humans are well
documented by medical and other professions and the popularity of wildlife-
related recreation evident in the extensive sale of nontechnical natural history
publications and the broadcasting of wildlife television programs. However,
it is also widely appreciated that wildlife-related recreation can produce a
whole host of undesirable side effects. Sustainable development requires that
in maximizing the benefits to visitors, these environmental costs must be
minimized.

Historically, public land has been the key to recreational opportunities in
the USA. However, there has been a tremendous increase in the demand
placed on this land in the last thirty years. An increasing shortage of recreational areas of public land, with subsequent overcrowding and poor maintenance, must lead to new opportunities for private landowners. Providing

opportunities for recreation away from public land may be particularly important in areas where there are high population densities but small estates of public land. For example, although 78% of the population live in the eastern states only 9% of public land in the United States is found there.

Wildlife-associated recreation is one of the most popular forms of outdoor recreation in the USA. In 1991 about 109 million Americans (some 56% of the adult population) participated in some sort of wildlife-related recreation (US Department of the Interior & US Department of Commerce, 1993). The ways in which people can derive benefit from wildlife are diverse. For clarification, recreational use of wildlife habitat is divided into:

consumptive uses, e.g. hunting, fishing; and
nonconsumptive uses, e.g. observing, identifying, photographing.

Surveys show that there is considerable overlap in the USA of participants in consumptive and nonconsumptive uses of wildlife. Expenditure for wildlife associated recreation in 1991 totalled $59.1 billion. Of this, $24 billion were spent on fishing, $12 billion on hunting, $18.1 billion on nonconsumptive activities and $5 billion on unspecified consumptive activities.

As populations expand in the USA, a growing recreational demand falls upon private lands. In order to encourage private landowners to provide for recreational use, a whole host of systems and incentives must be used, which will appeal to each individual's objectives. Costs to the private landowner of allowing recreational use of the land can be substantial. Owen *et al.* (1985) reported that litter, dumping, illegal firewood cutting, road damage, trespass and vandalism are serious problems. At present, certain institutions, such as weak liability statutes, may act as powerful disincentives for opening up private land to public use. Although a total of 1.2 billion acres of the USA is held in private ownership, estimates suggest that only 21% (255 million acres) is open for public use, either for free or with some sort of fee charged (Wright, 1989).

Provision of wildlife associated recreation must also take account of the potential environmental costs of that provision. It is often assumed that pursuits such as the viewing of wildlife result in little or no environmental damage to the animals or their habitat. Indeed, it is tempting to assume that participants might be expected to have a special regard for the welfare of the subject species and therefore pursue measures to minimize environmental impact. In reality, however, competition between viewers and a lack of understanding of the sensitivity of the habitat and the particular species to disturbance often results in damage to one or the other, or both. Edington & Edington (1986) commented that similar problems could arise in relation to the taking of wildlife through hunting and fishing: 'Although the modern hunter habitually safeguards the interests of his chosen quarry species, his single-minded attention to this aim frequently produces environmental disruption in other directions.'

Chapters 5 and 6 examine the combination of recreational enterprises with the conservation of natural areas. The chapters provide several case studies where recreational enterprises have been successfully integrated with the management of land for conservation of wildlife. There is a growing appreciation of the need to understand the ecological factors associated with recreation in order to reduce environmental damage. Therefore both chapters discuss the need to construct and operate positive management practices for minimizing damage to habitat and wildlife. In addition, it is recognized that the institutional arrangements surrounding the use of private land for recreation will affect the amount and type of provision supplied by landowners, and so the effectiveness of current institutional arrangements in providing private landowners with incentives for protecting wildlife habitat is evaluated.

This chapter specifically examines the use of wildlife habitat for fee-hunting and shows how the ability of private landowners to charge for access to hunting can act as a powerful incentive to protect wildlife habitat. Chapter 6 examines the nonconsumptive recreational use of wildlife and evaluates the potential for recovering the cost of wildlife conservation through 'wildlife watching' enterprises.

Consumptive use: fee-hunting

Fee-hunting for conservation

The American revolution expressed a resistance to the English system of hunting, whereby parliament controlled licenses to possess arms and private landowners controlled the right of access to wildlife:

'American tradition became opposing anything which restricted an individual's freedom to take game whenever or wherever he desired.' (Steinbach & Ramsey, 1988). At the turn of the century sparse human populations and abundant wildlife resources in the western states resulted in uncontrolled harvest. Native big game species were almost eliminated within 50 years of European settlement of the region. As a result control over wildlife was vested in the state. In 1646 the first hunting regulations were introduced in Rhode Island: a closed season was established for white tailed deer. This ordinance established a blueprint for hunting laws, which were adopted by most of the colonies before 1720. By 1878 Iowa had introduced the first set of regulations governing the amount of game taken – 'bag limits' (Trefethen, 1975). In these early days regulations were not introduced in response to recreational hunters. They were directed at controlling the removal of animals from the land and the taking of wildlife for meat and fur. Dunlap (1988) reported that the concept of hunting as a sport was only introduced into America from England in the early nineteenth century.

When measures are required to preserve species and numbers, the idea that hunting wildlife could help preserve or manage a species is an anomaly to some. Perhaps this is not surprising in the USA, where hunting by European colonists reduced populations of the passenger pigeon (*Ectopistes migratorius*) and heath hen (*Tympanchus cupido*) to the extent that they never recovered and, despite concentrated conservation efforts to save the latter, became extinct in 1914 and 1933 respectively. Hunting also caused significant decline in numbers of the whooping crane (*Grus americanus*), Eskimo curlew (*Numenius borealis*), plains bison (*Bison bison*) and pronghorned antelope (*Antilocapra americana*) (Trefethen, 1975; Halliday, 1978). The case against hunting today, however, is normally based on moral arguments (e.g. Singer, 1975) rather than ecological evidence. Generally, it is acknowledged that animal populations can be hunted on a sustained yield basis and that some populations require management to prevent population explosions. Along with the concept of the sporting hunter in the early nineteenth century came the sportsman's creed. The ideal of sportsmanship spread with the creation of sporting clubs, the most famous of which is probably the Boone and Crockett Club, formed by Theodore Roosevelt and George Bird Grinnell in 1887. The club contributed to sustainable hunting by establishing game refuges and encouraging game conservation (Trefethen, 1975). Increased regulation of sporting hunters was seen as a means of protecting the sport to the extent that in 1933 Aldo Leopold commented that the history of American management was 'until recently almost wholly a history of hunting controls' (Leopold, 1933).

Hunting regulations today govern licensing, methods of capture, limit seasons and impose bag limits on game. In addition to regulating hunting, the states' game departments have been responsible for transplanting game species back into the states. Since the turn of the century wildlife numbers have increased dramatically, despite the fact that wildlife habitat is declining. Interest in hunting game species continues to rise. In recent decisions, the Supreme Court has abolished the claim that wild animals are owned by the state but has ruled that they are held by the state in trust for its citizens. As such, the states have a right to regulate the hunting of wild animals but the perception that wildlife is a public resource is very strong.

Table 5.1. *Hunting expenditure in the USA by type, 1991*

Type	Species hunted	No. participants	Average annual spend
Big game	Bear, wild turkey, deer and elk	10.7 million (12 day average)	$476/person
Small game	Pheasant, rabbit, quail, grouse, prairie chicken and squirrels	7.6 million (10 day average)	$197/person
Migratory birds	Doves, ducks and geese	3.0 million (7 day average)	$233/person
Other animals	Coyote, fox, woodchuck and raccoon	1.4 million (14 day average)	$214/person

Source: US Department of the Interior, Fish and Wildlife Service & US Department of Commerce, Bureau of the Census (1993).

Economic demand for hunting

Hunting is very popular in the USA, with hunters making up around 8% of the adult population. Hunters spend an average of 17 days hunting each year and statistics concerning expenditure provide sound evidence of the demand for hunting. The average annual cost per hunter in the USA is $872 (USDI & USDC, 1993). In addition about 20% of the adult population fish, spending an average of 14 days fishing at an annual cost of around $675 each. Big-game hunting is the greatest contributor to hunting activity (Table 5.1) and the total expenditure on hunting may be broken down as shown in Table 5.2.

The breakdown between the different types of hunting reveals that big-game hunting accounts for 41% of all expenditure. Equipment accounts for the majority of hunting costs (42%), with trip-related expenses making up 28% of total expenditures. Breakdown of trip-related costs suggests that:

53% is spent on food and lodgings;
38% is spent on transportation; and
only 9% is spent on other trip-related expenses such as guide fees and
    rental equipment.

Approximately 83% of hunters hunt on private land. While some (15%) hunt exclusively on public land, 76% of all hunting days in 1991 took place on private land and 54% of all hunters restricted their hunting exclusively to privately owned land. A significant revelation of these statistics is the lack

Table 5.2. *Total hunting expenditure in the USA, 1991*

Trip-related expenditures		
food and lodging	$1.8	
transportation	$1.3	
other trip costs	$0.3	
Subtotal		$3.4 billion
Equipment expenditures		
hunting equipment	$3.3	
auxiliary equipment	$0.6	
special equipment	$1.2	
Subtotal		$5.2 billion
Other hunting expenditures		
magazines, membership dues and contributions	$0.2	
land leasing and ownership	$3.0	
licenses, stamps, tags and permits	$0.5	
Subtotal		$3.7 billion
Total hunting expenditures	$12.3 billion	

Source: Figures taken from US Department of the Interior, Fish and Wildlife Service & US Department of Commerce, Bureau of the Census (1993).

of correlation between the distribution of time and money spent on private land. It seems that little of the hunting-related expenditure is reaching the private landowner. While 76% of hunting time was spent on private lands, only 24% of hunting-related expenditures were associated with access to private land and most of this was for the purchase and lease of recreational land by individuals or private hunting clubs.

Supply of fee-hunting

In the USA, game mammals and birds are owned by the public at large. Historically, American hunters have had free access to both public and private land in the western states. After obtaining a license to hunt a particular species from the state game department, a hunter may hunt on public land or any private land which is not 'posted' to exclude hunting (see Fig 5.1). Although hunters should request the permission of private landowners, they are not required to do so by law in many states. An abundance of wildlife has meant that, until recently, hunters were generally able to gain permission from private landowners for no charge.

As the demand for hunting increases in the USA, there is likely to be a case for limiting the amount of hunting per parcel of land to maintain the quality and safety of the activity. A nationwide poll conducted by the

Fig. 5.1. 'No Hunting' sign, 'posted' on private land. Photograph: Victoria Edwards.

National Shooting Sports Foundation in 1986 revealed that hunters believe 'access to hunting land is the biggest problem of the sport today.' Indeed, 19% of the hunters surveyed cited access as a major hunting-related problem and 18% mentioned overcrowded hunting areas (National Shooting Sports Foundation, 1987).

Today, management of wildlife on public land and associated research is funded from the returns earned from an excise tax placed on guns and ammunition by the Pittman-Robertson Program and the Dingell Johnson Act of 1950. While private lands provide much of the opportunity for hunting in the USA, state agency budgets for wildlife are spent almost exclusively on public lands. Exclusion of access to hunting affords private landowners the opportunity to charge a fee and so simultaneously address the problem of the costs associated with providing wildlife habitat. It is a practice that has been common in several states for some time (for example, in Texas where 97% of the land is privately owned).

Hunters who are prepared to pay for access to hunt on private land may do so either through lease-hunting or fee-hunting. The simplest type of agreement is a basic access agreement with little or no services provided. As additional services are offered the agreement becomes more elaborate. Services might include accommodation, meals, guides, transport, packing and care of dogs and horses. Leases might be sold on the following basis:

1. Day hunting leases: access for hunting or recreation may be sold on a daily basis at a fixed rate per day. This usually provides the right to access with no recreational services.
2. Limited duration lease: this type of agreement generally gives the lessee rights for the entire legal hunting season on a specified game species, or 12-week intervals within a hunting season. Recreational services may or may not be provided.
3. Year round lease: the landowner leases the property on a yearly basis, generally to a party of hunters, a sports organization or even a recreation broker. The lease provisions generally include all hunting privileges and may also include other yearly recreational privileges.

McClelland *et al.* (1989) summarized the relative advantages and disadvantages of each (see Fig 5.2).

The fees charged vary considerably and depend upon the services provided, the perceived quality of the hunting, the trophy value of the animal/bird and hunting demand. Regional variations will exist according to local custom and demand. For example, in parts of Wisconsin, Canada geese are considered vermin by local farmers and hunting is encouraged to control them: access to hunting is usually free. However in southern states, where availability of waterfowl hunting may be more restricted, leases might reach $4 to $50/ acre (Soutiere, 1989). Tjaden (1989) reported that the average game hunting lease in southern states could range from $1 to $10 per acre. While in the past hunting clubs were only interested in leasing large tracts of land (1000 acres

Types	Advantages	Disadvantages
■ Day hunting lease	1. Generally yields higher average net returns per acre because of higher number of hunters per acre 2. Provides use for extra labor available during the hunting season 3. Guests can be invited to hunt without making any special arrangements with paying hunters 4. Allows for close control of harvests as a management tool	1. Requires considerable time during the hunting season since the hunter generally gives short notice to hunt 2. Most expensive arrangement to the landowner in terms of management cost 3. Income is generally not known in advance 4. Requires greater contact with people on a daily one-to-one basis
■ Limited duration lease	1. Requires less management in time and labor 2. Land owner generally knows hunters personally 3. Least expensive arrangement for landowner since contact is generally with one person or small group 4. Income is known in advance	1. On average, yields the lowest net returns per acre 2. Tends to leave little flexibility for regulating wildlife populations 3. Generally offers no influence or incentive for care and protection of wildlife during the off-season 4. Generally requires special provisions and arrangements for landowner–guest hunting
■ Year round lease	1. Landowner is generally removed from management responsibilities and is free to carry on usual activities 2. Hunting arrangements are generally provided by lessee 3. Depending on the arrangement, good hunter–landowner relationships may be established (not necessarily true with a broker lease arrangement) 4. Lease arrangements may be for several years 5. Income is known	1. Hunters or recreationists may be on land at undesirable times 2. Wildlife game species may be inadequately harvested 3. Landowner may not be aware of activities occurring on the property 4. Landowner may not have direct control over hunters once lease arrangement is signed 5. Landowner may be reluctant to increase lease price or change provisions of the arrangement 6. May yield lower net returns than some other type of management options

Fig. 5.2. Types of property leases for hunting with advantages and disadvantages Source: McClelland *et al.* (1989).

or more), smaller areas have become more attractive and recently areas as small as 100 acres have been leased

Generally, if the ranch is remote from urban populations then ancillary services, accommodation in particular, should be offered and the landowner has the opportunity to market a far more sophisticated product and so charge a higher price. This type of hunting is generally marketed on a set-fee basis for a specific number of days. It is likely to be more popular to urban hunters, who may not belong to a hunting club, and requires a greater degree of commitment and organization on the part of the landowner. Essentially, the landowner must take a much more active role in the hunting either providing or organizing guides, transport, accommodation, meals, packers, taxidermists, insurances, licenses, etc.

All of the ranches included in this research reported that marketing was essentially conveyed by word of mouth, with an occasional advertisement placed in a sporting magazine. In general, excess demand for the hunting currently exists. In terms of retaining client appeal, however, the quality of the hunting experience for the hunter remains as the most important criterion, with the price elasticity of demand appearing fairly low. The hunting experience is likely to depend on the success rate of the hunter, exclusivity of the site, safety, facilities and landscape. Belt & Vaughn (1988) confirmed this: 'Hunters want a *recreational experience* as much as or even more than taking home any deer, geese, quail or other game they have killed.' McClelland *et al.* (1989) recognized that the type of hunting client determined the need for a quality experience and smooth running operation, stating that, generally, hunting clients are:

well-informed about what services are available;
older individuals;
conservative;
affluent or at least above average in disposable income;
time and convenience conscious;
independent people who emphasize self-fulfillment; and
limited in the time available to them for recreational activities.

Case study: Deseret Land and Livestock Corp., Utah

Deseret Land and Livestock is a 200 000 acre landholding in northeastern Utah, owned by the Church of Jesus Christ of the Latter Day Saints (the 'Mormon Church'). Approximately 7500 acres of the privately owned lands are flood-irrigated and an additional acreage of federal land is leased for

(a)

(b)

Fig. 5.3. Deseret Land and Livestock Corporation Ranch, Ogden, Utah: (a) mountain
area; (b) plains. Photographs: Victoria Edwards.

grazing. The elevation of the ranch ranges from 6000 to over 8500 feet (Fig.
5.3). Vegetative types vary from salt-desert shrub species and sagebrush
grass, to mountain brush, quaking aspen grass and subalpine fir ecotypes.

The ranch has a diverse range of enterprises to match its diverse habitats.
The main business is a suckler cow and calf operation. From the 4200 suckler
cow herd, all of the heifer calves and from 50–100% of the steer calves are
kept to yearlings. Between 100–150 bulls run with the herd. In addition,

grazing is leased to neighboring sheep operators (supporting around 2000 sheep). Other enterprises include a seed operation for land reclamation projects and the wildlife project.

The ranch's wildlife department markets access rights to hunters who wish to pursue deer, elk, moose, prong-horned antelope and ducks. As a division of the entire corporation, the wildlife department is very important in terms of the ranch's economic viability. The wildlife enterprise generates from 25–50% of the ranch's net profit each year (depending upon the price achieved for the cattle). Perhaps more importantly, the wildlife enterprise has 'made us more balanced resource managers, and overall changed our attitude about wildlife. We have become more committed to the fact that wildlife is an important part of the land resource model: we want to preserve, protect and enhance this resource for its benefit, ours, and the public's.' (Simonds, 1988).

The ranch currently has game populations of (approximately): 2000 elk (summer), 5000 mule deer, 500 antelope and 50–100 moose. The elk have become much stronger in number since the inception of wildlife management at Deseret in 1976. Numbers have increased from 350 to the current 2000 and reproductive rates are now 65% (compared with 55% on neighbouring land and 17% in Yellowstone National Park) (Simonds, 1988). Surveys on the ranch of nongame species show that there is an increasing presence of squirrels, jack rabbits, beaver, mink, bobcat, mountain lion, cougar and black bear. Raptor nesting sites (including bald and golden eagles) are inventoried, as are populations of migratory birds such as cormorants, egrets, terns and sage grouse.

At present, the ranch's wildlife department currently derives all of its income from hunting fees. However, it would like to branch out into nonconsumptive wildlife use of the land. Approximately 1500 vehicles pass the entrance to the ranch each day. In addition it is within two hours' drive of Salt Lake City (Utah's capital), with a population of around 200 000. Many people already stop at the ranch boundary to view the elk grazing in winter. A joint program for developing a wildlife viewing area has been planned with the Bureau of Land Management and the State of Utah, whereby the Bureau will provide a parking lot, the State the signs and the ranch the fences and patrol. In addition, the ranch is planning to operate wildlife workshops and is investigating the possibility of opening a visitor center. This would be consistent with the Church's objectives for the ranch: its current five year plan states that 'we will allow greater access in a controlled, workable manner', and sets a task to 'pursue in a formal manner educational/recreational activities that would be adaptable to ranch resources, powerful to participants, and very positive politically.'

Benefits and costs of fee-hunting

Managing a farm for fee-hunting can result in net economic benefits for a landowner, rather than net costs. Fee-hunting enables private landowners to recoup some of the costs involved in conserving wildlife habitat on their land. Costs may include damage to domestic livestock feed and fencing, as well as the opportunity cost of using natural wildlife habitat for alternative uses. In 1986, a study of big game on private land in Utah reported that the average estimated losses or damage to farmers by big game animals feeding on cropland were: $9.35/acre for elk and $6.15/acre for deer. Figures for grassland were $3.75/acre for elk and $1.22/acre for deer (Kwong, 1987). The Conservation Reserve Program (CRP), introduced in the 1985 Farm Bill, was designed to set aside up to 45 million acres of highly erodable, marginal cropland for five years. Landowners may retire the cropland to trees, permanent wildlife habitat, permanent grasses and legumes or combinations. While the retired croplands may not be grazed or harvested, they may be used in a commercial manner through leasing for hunting. The program, therefore, offers the opportunity to establish a wildlife area on a farm and yet still derive some income from the land.

In addition to paying for access to the land, fee-hunters may pay for associated services, such as a guide, transport (four wheel drive or horseback), accommodation and game preparation after the hunt. Fee structures are dependent upon the type of hunting offered and the related accommodation and services. Most ranches with a comprehensive fee hunting program also have an attractive guest house on site. The prices that hunters are currently prepared to pay at Deseret Land are listed in Table 5.3 together with the number of hunters for 1994.

In addition to charging for hunting, Deseret grants 10 antelope hunters free access to the ranch from a public draw. The ranch has an agreement with the State Game Department that the antelope season will be extended from one week to two months as a result of this arrangement. Several hundred hunters are also allowed free access to the ranch to shoot doe deer and a further 160 hunters to shoot doe antelope.

It must be noted that, in order to derive income from fee hunting, landowners must be prepared to pay the associated labor and feed costs of maintaining a game population. Deseret Land employs one full-time and two part-time wildlife biologists and several graduate students (Fig. 5.4). Guides are employed on a weekly basis at a cost of approximately $1000 – $1200 per week. The ranch also employs a cook and packer for each hunt. Revenue

Table 5.3. *Deseret Land – hunting fees 1994*

Type of hunt	Price ($) per person	No. of hunters 1994
Elk		
5 day guided hunt – trophy bull	8000	16
5 day guided hunt – 5 point bull	4200	12
5 day guided archery – bull	5000	4
[a]unguided hunt – trophy	3000	6
[a]unguided hunt – 5 point bull	1500	8+
[a]unguided archery hunt – bull	1500	4
hunting access – cow	100	300
Deer		
5 day guided hunt – stag	4000	12
hunting access (11 day season)	1300	110
archery access (2–3 wk season)	700	7
[a]Unguided hunt – stag	1750	22
Antelope		
guided	2200	20
Moose		
5 day guided hunt	5000	4
Ducks		
club leased (1 month season)	500	5

[a] Where the landowner grants permission for the hunt to a private guide who then sells the hunting rights.
Source: Deseret Land and Livestock Corporation, Utah.

Fig. 5.4. The author with Rick Danver, wildlife biologist at the Deseret Ranch. Photograph: Greg Simonds.

from hunting must also cover the cost of the hunter's food, fuel and the maintenance of the guest house.

Although hunting is available free on all public land and some private land, hunters are prepared to pay for access to private land when they perceive that the quality of the hunting experience is increased. While demand for fee hunting may be strong where access to public land is scarce, such as in Texas, it may also be strong where public land hunting has become overcrowded and often dangerous. Most big game species have a relatively small designated hunting season (on average around 10 days). Concentration of hunters on public land can be considerably high during these periods and there have been many reported accidents amongst hunters. Limiting the number of hunters in any one area and controlling the movements of each hunter can greatly enhance the safety aspects of the sport.

In addition to enhancing safety, controlled private hunting enhances the quality of the hunt. On the Deseret ranch, success rates for hunters of deer and elk range from 75–100%, compared to the 20–30% success rates on neighboring public land. The ranch limits access to approximately 200 deer permits each year now, compared with the 3000 hunters who used the land before hunting was restricted.

While generating income for the private landowner, fee-hunting also substantially improves wildlife management. In most parts of the USA natural predators, such as the bear, wolf, mountain lion, coyote and bobcat have been eradicated because of their damage to domestic livestock. As a result other game species, such as deer and elk, flourished and began competing for valuable grazing or causing damage to farm crops. Destruction of habitat for wildlife was often seen as a solution to reducing the numbers of big-game species present on a farm. Fee-hunting provides the private landowner with the incentive to maintain wildlife habitats, while at the same time keeping the game species in check and preventing them from destroying natural habitats for rare plants and other animals: 'Numerous case studies on corporate lands have reported that lands open to unrestricted public access tend to be the most abused and have low wildlife populations.' (Owen, 1989).

At Deseret, they believe that the wildlife department makes the best use of nature for a variety of purposes. Beaver have been introduced into areas that have suffered badly from soil erosion and, in the areas where the beaver have not stayed, the wildlife biologists have mimicked the beaver by building small dams to help catch silt and stabilize stream banks. Willow shoots have also been planted for reestablishing willow in riparian areas.

Fig. 5.5. Dye Creek Preserve *before* wetlands restoration. Photograph: Chuck Harrison.

Case study: Dye Creek Preserve, California

Dye Creek Preserve is a 37 000 acre ranch in northern California. The ranch comprises some lowland grassland areas but mostly high open range, spotted with oak and divided by rugged canyons. The ranch has had a chequered history of ownership. The property was acquired by a state bank when the previous owner defaulted on a loan. Since the bank owed the State of California around $150 million, the State acquired the ranch, along with several other properties, as part payment. Dye Creek was then put under a trust, which is now managed by two other banks who, in turn, have leased the property to The Nature Conservancy for 25 years (W. Long, personal communication 1994).

A fee-hunting and recreation program had been run on the ranch for the past 25 years. In order to finance the management of the land, the Conservancy has continued to lease the land to the fee-hunting management company (Multiple Use Managers, Inc.) and leases grazing rights to the previous cattle rancher.

Dye Creek provides a good example of income opportunity from the conservation of waterfowl. Work is carried out to maintain lakes that were established in the 1960s to attract migratory waterfowl. For example, goose platforms have been built to enable geese to nest away from predators, boxes have been erected for wood ducks and levees are maintained in order to retain water levels (Figs. 5.5, 5.6). Spring numbers of Canada geese have risen from 12 birds in 1986 to 75–100 birds in 1991.

Less active management is needed to encourage the healthy production of other game species. Feral pigs, deer, wild turkeys, doves and quail are managed by restricting hunting on the land and controlling population numbers.

Fig. 5.6. Dye Creek Preserve *after* wetlands restoration. Photograph: Chuck Harrison.

Hunting is leased through permit or guided shoots. Dye Creek is prime hunting ground for the Eastern Tehama deer herd – the largest deer herd in the State of California. With a combination of low-elevation oak and migrating canyons, the ranch serves as good winter habitat for the herd. The acorns of the blue oaks, in the upland area of the ranch, provide a valuable source of winter nutrients for deer. Success rates for deer hunters at Dye Creek are around 85%, compared with 10% for the state's average (Fig. 5.7). In addition, Dye Creek offers hunting of wild boars, wild turkey, waterfowl, dove, quail and fishing for trout and bass. The success rate for wild boar is recorded at 98%. Table 5.4 summarizes the availability and price of hunting at Dye Creek.

Preserving the habitat of the upland areas for game wildlife has also helped protect nongame species at Dye Creek. For example, black bear, mountain lion, otter, beaver, raccoon and skunk inhabit the ranch, together with bald eagles, egrets, blue herons and turkey vultures. The land is steeped in cultural and natural history and the wildlife managers of Dye Creek have recently begun to open up the ranch for more nonconsumptive use of wildlife, such as 'Guided Photography Safaris'. A small lodge in the valley of the preserve is used to house guests and can be hired for business retreats or family holidays. The Nature Conservancy already operates field trips on the property for its members.

Case study: Mount Taylor Game Ranch, New Mexico

Mount Taylor Ranch is a 7000 acre private game ranch of high mountain land in New Mexico. The property is heavily timbered and has an elevation

Fig. 5.7. Successful hunter with black-tailed deer at Dye Creek Preserve. Photograph: Chuck Harrison.

from 5000 to almost 10 000 feet (Fig. 5.8). The ranch, which is also managed by Multiple Use Managers Inc., has a very attractive guest lodge. The lodge, with its impressive stone and timber interior, has several large reception rooms and 18 bedrooms, each with ensuite facilities. It is situated at 9000 feet and is surrounded by several trout fishing ponds (Fig. 5.9).

Wildlife on the ranch includes mule deer, elk and black bear. Hunting for elk is priced at $7000 for four days (1993), including accommodation, meals, guides and transportation from Albuquerque airport. The bugle season (and best hunting season) is during the first three weeks of September. Transport around the ranch is in a four-wheel drive vehicle or on horseback or foot. Fees are also charged for the hunting of mule deer and Himalayan tahr. Since Mount Taylor is a 'game ranch', the landowners have been allowed to import

Table 5.4. *Dye Creek Preserve – hunting fees 1993*

Hunting	Season	Cost
deer	End Oct.–mid Nov.	$7000 per individual or $1100 for a 3 day guided hunt (incl. transport, meals and accommodation)
waterfowl, dove, quail, trout and bass	Regular state season	$1000 per year (limited to 25 members)
wild boar	Nov.–April	$600 (with additional trophy fee of $200 for tusks of 2 in or longer)
wild turkey		$300/day (2 day minimum)

Source: Multiple Land Use Managers Inc., California.

Fig. 5.8. Mount Taylor Ranch, New Mexico. Photograph: Victoria Edwards.

exotic species onto the land subject to them being fenced in. The entire property is fenced in two 3500 acre parcels: the maximum permissible area for a game ranch according to New Mexico State laws. A three-day tahr hunt is charged at $1500, with an additional $1000 trophy fee. The majority of hunters who stay at the ranch are from out-of-state and tend to enjoy a higher income than the USA median (around $25 000 p.a.). Marketing is by word-of-mouth and through Multiple Use Manager's quarterly newsletter. The

Fig. 5.9. Trophy elk grazing at Mount Taylor Ranch. Photograph: Victoria Edwards.

improvement of wildlife on the ranch is actively pursued through a feeding policy. In this respect, Mount Taylor comprises very much a game ranch, managed for hunting income, rather than a ranch which happens to support wildlife.

Hunting and conservation: potential problems

Many recreational uses are directly compatible with the maintenance and enhancement of wildlife areas. This chapter has illustrated how consumptive use of wildlife on private land in the USA can help to generate income to assist in the management of natural habitat. Fee-hunting provides an income source and strong incentive to improve areas of habitat for game species. Often, nongame species will benefit from such habitat improvement programs.

However, there remain a number of problems with managing fee-hunting on private land, which must be overcome if wildlife and wildlife habitat are to benefit from the enterprises. First, there is the need to manage hunting of the game species on a sustained yield basis. While trophy bucks and bulls attract a premium from fee-hunters, private landowners must be careful to manage their elk and deer populations in a way that will retain the correct balance of males and females and will support *long-term* production of trophy animals. Deseret Land tackles this problem in several ways. First, by insisting that all unguided deer hunters must shoot a doe, regardless of whether they

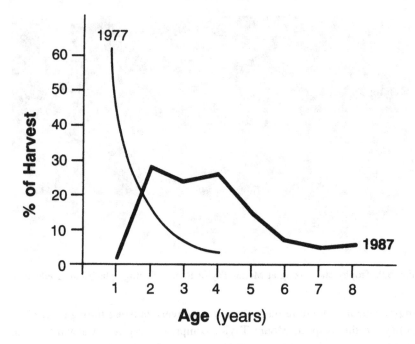

Fig. 5.10. Deseret Land and Livestock Corporation – harvest of deer by age group 1977 & 1987. Source: Simonds (1988).

shoot a buck. Second, it sells two types of controlled guided elk hunts: trophy elk (6 points or more) and 'bull management' hunts (where bulls of 5 points, which are unlikely ever to reach trophy stage, are culled). Bulls that have the potential to reach trophy stage are thus retained. Figure 5.10 shows how the quality of their herd improved in 10 years of positive management. The graph shows that the age structure of harvested deer at the beginning of the program, in 1977, has a much less broad base than in 1987, when a more even representation of age range was harvested.

In order to sustain populations of game over time (a) animals must be kept free from disturbance during the breeding season and (b) the level of harvesting must take into account annual variations in reproductive rates, disease, nutritional problems, predation level, and other causes of mortality. Both can be achieved through the positive management of game on private land coupled with the strict control of hunting.

Apart from the obvious danger of failing to manage hunting on a sustained yield basis, the introduction of exotic species and the persecution of predators can reduce the wildlife benefits of fee-hunting.

Introduction of exotic species

Enthusiasm for introducing exotic game species into new habitats can cause serious depletion and even destruction of native species. Damage is generally brought about by either direct attacks on native species from the exotic newcomers, or competition for the same resources. New Zealand provides horrifying testimony to the damage that can be caused by the introduction of exotic species for hunting purposes. Prior to the introduction of exotic species, the native fauna of New Zealand included only two species of land mammals, both species of bats: the long-tailed bat (*Chalinolobus tuberculatus*) and the short-tailed bat (*Mystacina tuberculata*). The Polynesian rat, 'Kiore' (*Rattus exulans*) and the native dog 'Kuri' were introduced by Maori settlers. The European settlers, in the pursuit of sport, were responsible for the introduction of the rabbit (*Orcytolagus cuniculus*), the hare (*Lepus europaeus*), the chamois (*Rupicapra rupicapra*), the Himalayan tahr (*Hemitragus jemlahicus*), and seven species of deer, including fallow (*Dama dama*), red (*Cervus elaphus*), rusa (*Cervus timoriensus*), sambar (*Cervus uniclolor*), sika (*Cervus nippon*), white-tailed (*Odocoileus virginianus*) and the wapiti (*Cervus canadensis*). Introduction of animals to the wild through the release of game species for hunting was added to by the escape and establishment of feral populations of cats, dogs, cattle, horses, pigs, goats and sheep. The most successful of these in establishing wild populations were the goats (*Capra hircus*) and pigs (*Sus scrofa*) which are hunted today.

The introduction of exotic species into New Zealand was carried out without a proper understanding of, or appreciation for, the effect that they would have on the grasslands and forests (which had never been browsed) and the birdlife which, in the absence of predators and disturbing presences, included many ground-nesting and flightless species. Barnett (1985) reported that 'the consequences have been disastrous and the realization that deer in particular were causing severe damage to native forest and so destroying wildlife habitats came too late to save many forests from destruction and prevent dramatic erosion of high country where tussock tops had been grazed clean.' Ferrets (*Mustela putorius*), stoats (*Mustela erminea*) and weasels (*Mustela nivalis*) were released by settlers in an effort to control huge populations of rabbits and hares that had arisen in the absence of any predators. The mustelids, together with introduced rats, cats and mice, have been largely responsible for the depredation of native birds: taking eggs and killing fledglings. Barnett (1985) summarized, 'it is an indictment of European settlement that today some ten percent of the world's endangered bird species should be from the

New Zealand region.' The New Zealand lesson is certainly one that the United States might take heed of.

Control of predators

A second cause of wildlife deprivation through hunting stems from the attempts of hunters and wildlife managers to control predation. If hunting enterprises perceive that they are competing with natural predators for the targeted species, they may attempt to control predators and so increase the hunters' yields. In particular, attempts may be made to eradicate natural predators to elk, deer and antelope. At present, game laws dictate that animals such as mountain lions, bears and bobcats must not be shot or trapped. If these animals are destroying domestic livestock then special permission may be obtained from the game departments to remove such animals. Unfortunately, a certain amount of illegal hunting of these animals does take place and there is debate over whether this has a significant effect on game yields. For example, Connolly (1980) identified a large number of situations where the size of the hunted population appeared to be limited by factors other than predators.

The dubious merits of predator control may, therefore, deter game mangers from interference. However, Edington & Edington (1986) cited three situations where predator control may be justified: (a) the reintroduction of native species for conservation purposes, (b) high winter mortality amongst game stock without a commensurate effect on their predators and (c) where the rearing of game species in game ranches increases vulnerability to predator attack. Reference to the third situation is, for wildlife purposes, somewhat disturbing. The authors were in no way advocating predator control: indeed they commented that in most game-rearing operations 'there is scope for introducing features which minimize loses due to predators, rather than adopting a policy of predator control.' However, citing the problems of predators in game rearing forces acknowledgment that the dangers of importing exotic species for game ranching may include not only the danger of such species competing directly with native species, but also that the managers responsible for their keep will favor conservation of the exotic species over that of native species.

To a less definite extent, landowners who manage for game species may simply not be encouraged to manage habitats in a way which will attract nongame species, particularly those that compete with game species for resources. This may be true where landowners directly feed game species to encourage them to remain on the land, rather than maintain natural habitat in sufficient quantity to enable a diversity of species to be supported. Generally, when the latter practice is used, nongame species will benefit along with the game species

Institutional reform

In addition to problems at the management level of a fee-hunting enterprise, there are a number of imperfections in the market for fee-hunting that need to be addressed.

Public perceptions

First, public perceptions, particularly in the western states, may hinder the acceptance of private fee-hunting: American hunters have become used to the right to hunt without charge. Now that private landowners have begun to exercise their property rights more forcibly, some hunters feel that they are being denied their right to hunt freely. As a consequence, there is some political pressure to stop fee-hunting on private land. One of the arguments used against fee-hunting is that it is turning into 'the rich man's sport it is in Britain.' Ironically, expenditure figures on hunting suggest that it has already reached that stage in the USA, but that only a small proportion of total expenditure is being put back into the land, with the majority being expended on equipment and trip-related expenses.

The public/private interface

Second, problems arise when hunting on private land is inter-mixed with public land. The fact that wildlife herds will mix over public and private land may inhibit the opportunities of the landowners to develop private wildlife management programs. Often the wildlife winter on private land (which tends to be in more lowland areas) and so compete for the farm's valuable winter forage crops. Summers, however, might be spent on higher, public land. If the game wildlife remain on this land during the hunting season there is little incentive for the private landowner to attempt to operate a fee-hunting enterprise.

The proximity of public land may also exacerbate the problem of trespass on private land. If a hunter shoots game on private land without the permission of the landowner he may be charged for *trespass*. However, provided the hunter has a licence to shoot that particular species then he or she will not be charged for *poaching*, since wildlife remains public property. Deseret

Land has a history of trespass problems and now employs a helicopter pilot and team of 'vacationing' policemen during the summer to patrol the ranch: each are paid on an hourly basis. States with a longer history of private fee-hunting and greater areas of private land may expect less problems from trespassers:

In all of Texas there is no acre of land on which an individual hunter can trespass in order to shoot a deer. Land privately owned is sacrosanct. Indeed, in Texas the verb to trespass is identical with the verb to commit suicide, for it is tacitly understood that any red-blooded Texan is entitled to shoot a trespasser.

Michener, 1985: 506

Landowner's liability

Third, attention must be paid to the litigation problems private landowners face in allowing access to their land for recreational activities. The extent of a landowner's liability towards a person who enters onto his or her holding depends upon the state laws. Different statutes determine the amount of care or duty that a landowner must use to protect a visitor and the extent of liability. Generally, the status of visitors can be defined as 'trespasser', 'licensee' or 'invitee'. An individual using the land for recreational purposes, if entering with the permission of the landowner but not paying a fee, would be classified as a 'licensee' (social guests are classified as licensees, not as invitees, even though they may have been 'invited'). Most clients of hunting enterprises, or of any fee-paying recreational use, would probably be classified as 'invitees'. Of the three classifications, the classification 'invitee' imposes the most onerous burden of care and liability on a landowner. An invitee is a person who is invited or permitted to enter or remain on land for a purpose of the landowner. Some courts require that the invitee affords some economic benefit to the landowner, others only require that there is an implied representation that care has been exercised to make the land safe for the visitor. The landowner is required to exercise reasonable care to make the land safe for the invitee or to warn the invitee of dangerous conditions or activities, which the landowner knows of, or those which he or she should discover with reasonable care. In an attempt to protect landowners from the threats of liability and to encourage access, the Council of State Governments drafted a model 'Recreational Use Statute' in 1965. All states, with the exception of North Carolina and Alabama, have mandated similar legislation, limiting private landowners' liability for injuries to persons using their land for recreational purposes. However, while most statutes insulate landowners from liability as long as access is permitted without charge, exemption generally does not apply if 'consideration' is required for use of the land. Some states have modified their recreational-use statutes to allow a landowner some flexibility in accepting funds to offset the costs and expenses of making land available (such as, Texas, Arkansas and Wisconsin). Several other states have considered amending recreational-use statutes to extend limited liability immunity to landowners who adopt a fee policy. Certainly, enactment of laws that, at the very least, allowed landowners to cover the costs of providing public access to natural areas might act as an incentive to encourage such provision. Further, statutes might be changed in order to clarify any ambiguities affecting the statute's coverage. At present, neither the coverage offered to landowners nor the statute's application in particular circumstances is clear. Therefore the uncertainty that surrounds the landowner's liability may act as a disincentive when landowners are deciding whether to open their land for recreational use.

The policy underlying the 'consideration' exception to recreational use statutes is to retain liability of the landowner when use of an area is granted for economic benefit. In such situations, the ability to derive income from the site is considered a sufficient incentive to encourage landowners to allow access: landowners may purchase liability insurance and spread the cost of accidents among all users of the land. As a result of the uncertainty surrounding the exact nature of the limited liability afforded by the state recreational-use statutes, all landowners with recreational enterprises on their land are advised to take out liability insurance. In addition, incorporation of the recreational business may be one way to lessen the burden of the landowner's liability. A corporation is considered a 'legal person' under state and corporation laws. This legal procedure generally limits the liability claims to the value of corporate assets. Careful selection of clients may also help to reduce the landowner's exposure to claims. On all of the fee-hunting ranches and wildlife preserves visited, client behavior (in safety terms) was a major factor of management, and staff were strict to point out the rules governing client behavior and any dangers that may be encountered.

Summary and conclusion

This chapter has shown how the incorporation of a fee-hunting enterprise on private land can help to defray the costs of protecting land for wildlife conservation. It has identified, however, that there are several potential problems with managing for game species, which must be taken into account in the overall management of the land.

There is no reason why recreational hunting should not be managed so as to minimize the environmental damage to habitat and wildlife species alike. The ability of private landowners to restrict access to their land and to control hunting behavior on their land can be used as a powerful tool in such positive management of hunting for wildlife compatibility. Private landowners are able to establish and enforce codes of practice which dictate what may be hunted, where and when hunting may take place (within hunting seasons) and how the subject may be killed and by whom. The ability to properly control the hunt in this manner can result in a safer, more enjoyable experience for the hunter, as well as facilitating wildlife management for the proprietor.

6

Watchable wildlife

Broadly speaking, a piece of scenery snapped by a dozen
tourist cameras daily is not physically impaired thereby, nor
does it suffer if photographed a hundred times. The camera
industry is one of the few innocuous parasites on wild nature.
Aldo Leopold (Conservation Esthetic)

Introduction

The United States has national and state park systems extending over 85
million acres. The original intention of this form of national land reser-
vation was to prevent large areas from being broken down into piecemeal
private estates during the great 'land grab' of the nineteenth century. Since
preservation, however, the areas have become popular with recreational
visitors. By 1970, the National Parks were receiving 172 million visitors
each year. During the 1980s visitors increased to between 300 and 350
million annually. The more widely distributed and more readily accessible
state parks have recorded as many as 660 million visits annually over
their relatively smaller total area of 10 million acres (Paterson, 1989).
Whelan (1991) reported that in Minnesota, visits to the state's 64 parks
increased from 6 million to 10 million in just three years.

Increases in personal mobility, leisure time and disposable income seem
set to increase the popularity of natural areas for recreation and amenity.
To many people, one of the most appealing characteristics of outdoor
recreation is the feeling of solitude and calm felt in an area of scenic
beauty. Ironically, the 'honeypot' effect of National Parks often ensures
that such solitude is rarely found. Yellowstone National Park receives
around two million visits a year: 'most visitors in the season drive the
park's roads in a convoy of cars which grinds to a halt every time a
motorist sights a bear' (Paterson, 1989). In 1972 the Council on Environmental Quality reported on environmental problems facing US National
Parks, citing heavy visitor use and poorly planned developments within
and adjacent to parks that were eroding the quality of visitor experiences
and affecting the parks' ecosystems. The feeding of animals and harass-
ment of wildlife, compounded by overcrowded conditions, were cited as

99

two particular problems in the parks that demanded changes in park management (Council on Environmental Quality, 1972).

Despite attempts to modify the behavior of visitors and so lessen their impact on natural areas, it is unlikely that America's National Park system, nor its state parks, will be able to satisfy future recreational demands. In particular, overcrowding of the parks will prevent them from supplying the one product that their visitors demand, the opportunity to observe and enjoy wildlife in their natural habitat: 'increased visitation to parks nationwide has resulted in more roads, more parking lots, and more concessions built in the protected areas, frequently decreasing the aesthetic value of the park' (Whelan, 1991: 13). As the demand for wildlife related recreation increases, opportunities exist for private landowners to combine natural area protection and management with recreational enterprises.

Demand

Hunting and fishing activity has been studied extensively in the USA, largely due to availability of licence records. However, it was not until 1975 that a national survey was conducted to examine other ways in which people benefit from wildlife. In 1985, a similar survey sought state-level information about those who observe, photograph and feed wildlife. Surveys conducted since 1975 indicate that nonconsumptive wildlife recreation in the USA has increased greatly in recent years. Results of the 1991 national survey showed that observing, photographing and feeding fish and wildlife provided enjoyment for 76.1 million American adults (39% of the adult population). Figures include nearly 30 million people who took trips of at least one mile from their homes for the primary purpose of engaging in nonconsumptive wildlife activities (some 16% of the adult population) and a further 73.9 million people (38% of the adult population) who took a primary interest in wildlife while around their homes. The 76.1 million participants spent over $18.1 billion on their activities in 1991: an average of $238 each (USDI & USDC, 1993). Expenditure is broken down in Figs. 6.1 and 6.2.

Three nonresidential activities were monitored by the survey: observing, photographing and feeding wildlife. The most common activity enjoyed by the participants was observing wildlife (28.8 million), with photography second (14.2 million) and 'feeding' wildlife a surprisingly high third (13.3 million) (USDI & USDC, 1993).

Unlike hunting and fishing, which are dominated by male participants (92% and 72% respectively), almost equal numbers of men and women

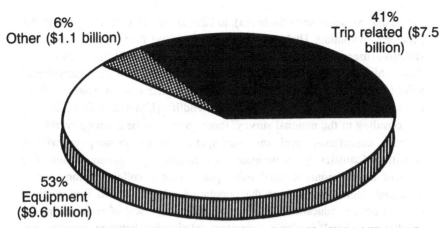

Fig. 6.1. Breakdown of total nonconsumptive expenditure 1991 (£18.1 billion total). Source: Figures taken from US Department of Interior, Fish and Wildlife Service & US Department of Commerce, Bureau of Census (1993).

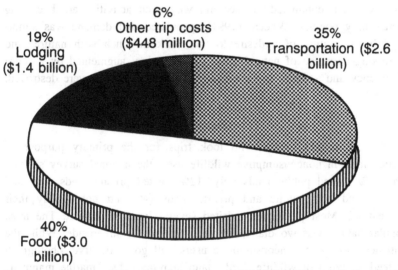

Fig. 6.2. Breakdown of trip related expenditure 1991 ($7.5 billion total). Source: US Department of Interior, Fish and Wildlife Service & US Department of Commerce, Bureaeu of Census (1993).

participate in nonconsumptive use of wildlife. Nonconsumptive wildlife use away from home was enjoyed by a substantial number of people from both rural and urban areas. Distribution of consumers was also fairly evenly spread across the regions, ranging from a participation rate of 12% of the adult population in the 'West South Central' region (Texas,

Louisiana, Arkansas and Oklahoma), to 22% in the 'Mountain' region (Idaho, Montana, Wyoming, Utah, Colorado, Arizona and New Mexico). Nonconsumptive users were more likely to spend days outside their own states than fishing and hunting wildlife users: although 68% of all nonresidential wildlife users spent time enjoying wildlife in their resident states, 32% of users travelled to other states to enjoy wildlife (USDI & USDC, 1993).

According to the national survey, there appears to be a strong correlation between educational level and participation in nonconsumptive wildlife activities. Statistics on participants' educational qualifications show that 16% of all the nonresidential participants have a college diploma or an advanced degree. Indeed, of the population who have had five years or more of college education, 28% participate in some sort of nonconsumptive wildlife use. There is also a consistent relationship between income level and participation rates (USDI & USDC, 1993).

While there is good information on the amount of watchable wildlife activities taking place and associated expenditure, no comprehensive studies have been conducted to ascertain why such activities are becoming increasingly popular. Whelan (1991) speculated that demand was stimulated by a mixture of a desire to 'get back in touch with nature' and increasing coverage of natural areas on television, augmented by a sense of urgency and a desire to visit natural areas before they were destroyed.

Supply

Of the 30 million people who took trips for the primary purpose of participating in nonconsumptive wildlife uses, the national survey showed that 51% visited public lands only, 12% visited private lands only and 33% visited both public and private lands (4% did not specify their destination). Most participants visited several types of habitat. The most popular habitat was woodland, followed by lake or streamside. While the statistics show that nonconsumptive users will go to observe, photograph or feed all types of wildlife (birds, land mammals, fish, marine mammals and others), birds, followed by land mammals are the most popular wildlife. Often game species, particularly rabbits, waterfowl and deer (more readily associated with consumptive uses) are very important in providing the opportunity for such activities.

Historically, state agencies have focused wildlife management primarily or entirely on game species, partly as a direct result of the funding process. Most of the funds for the management of fish and wildlife have been derived from the sale of hunting and fishing licenses and from excise taxes collected on selected pieces of hunting and fishing equipment. In a 1980 government survey of people's opinion about nongame funding, strong support was expressed for the general concept of funding the conservation of nongame species (Shaw & Mangun, 1984). However, when asked to rate systems of funding, most people favored systems that were voluntary. This is entirely consistent with the tradition of conservation of nongame wildlife in the USA, where private organizations have dominated the conservation market for the last 60 years.

Clearly there is an identifiable market, comprising people who wish to participate in nonconsumptive use of wildlife. For the private landowner the demands of these participants may bring both benefits and costs. In a survey conducted by Bromley (1989) of wildlife extension officers working with landowners, no officers reported landowners who gained income from nonconsumptive recreational use of wildlife. Activities related to wildlife that could generate income include bird watching, nature hikes and the harvesting of renewable resources, for example berry picking, mushroom hunting, shrub and fern digging and gathering of plants. In addition, more traditional recreational activities that may be compatible with wildlife conservation include horseback trail rides, backpacking, camping, picnicking and fishing.

The lack of response from landowners to increasing demand for wildlife related recreation and access to private lands might stem from a recognition that allowing public access to private land can be expensive. A survey by Kansas State University in 1988/89 showed that 57% of landowners surveyed had encountered problems with allowing public access to their land. The public access problems (which included trespassing, livestock/crop/property damage, litter/garbage, gates left open, tampering with machinery, starting fires, creation of erosion problems and fences cut) resulted in financial losses of up to $10 000 (Hildebrandt, 1989). Landowners with special features on their land, in terms of wildlife and habitat, are unlikely to advertise their existence if they perceive that public interest will result in increased costs. Not only have landowners to fear the costs associated with visitor pressure on the land, as above, but they may also be wary of the extra costs of land management that might be imposed if the public press for conservation of a particular species or habitat. However, finding ways of combining conservation with fee-paying recreational enterprises, enables landowners to reap financial benefits from wildlife protection and so provides a strong incentive for sound management of natural areas. Values associated with the activities described above – observing, photographing and feeding wildlife – are essentially 'use' values and are therefore largely excludable. Landowners who are able to *exclude* persons from enjoying wildlife on their land may derive income from allowing access to the wildlife and from providing associated services. Individuals might pay not only for access, for example on a per capita basis or through some membership scheme, but also for associated services. These may include:

interpretation (e.g. published material, exhibits or a naturalist guide);
workshops (e.g. photography, observation skills, etc.);
facilities (e.g. food, accommodation, transport); and
special purchases, in the form of gifts at the site.

In the western states, the most common form of obtaining income from nonconsumptive use of wildlife is to combine the fee-paying activities with a guest ranch. Guest Ranches evolved in the USA in the late nineteenth century. People from the eastern states, and some tourists from Europe, wanted to experience the 'wild west' and landowners were keen to find substitutes for their falling cattle economy. Guests were expected to participate in the activities on the ranch and were delighted in a sense of accomplishment from their duties. Today, the guests want to combine the reality and participation of a ranching enterprise with present day comfort (a 'real wilderness simulation') and ranches are beginning to realize that nature tourism may help to diversify their income source. In Wyoming, Montana and Idaho, ranch and farm hospitality enterprises (excluding those offering hunting opportunities only) increased from a mere handful in 1985 to between 70–90 in 1991 and are estimated to generate at least $750 million per annum (Bryan, 1991). Consistent with fee-hunting, it seems that the quality of the experience will remain an important marketing tool for nonconsumptive wildlife enterprises. This may be particularly true as people find more activities they wish to pursue in their limited leisure time: 'quality is the watchword for the 90s and beyond' (President's Commission, 1987).

Case study: Pine Butte Preserve, Montana

The State of Montana is rich in scenic beauty and natural resources. Western Montana is dominated by the Rocky Mountains ('Montana' is derived from the Spanish word for mountain), while isolated mountain ranges and high plains cover the eastern and central parts of the state. The state currently attracts nearly four million visitors a year and it is estimated that recreation will overtake agriculture as the major economy in the next couple of years. Over three quarters of all state tourists to Montana reportedly visit national parks while vacationing, and over 90% of those rate Montana as 'good' or 'excellent' in terms of outdoor recreation (Brock *et al.*, 1990).

Pine Butte Swamp Preserve is located along the eastern front of the Rocky Mountains in west-central Montana. It is 27 miles west of Choteau and 60 miles southeast of Glacier National Park. The 18 000 acre reserve is owned by The Nature Conservancy, a private nonprofit conservation organization. The reserve was acquired by The Nature Conservancy in 1978. Pine Butte itself (a 'butte' being a steep-sided hill) was formed during the last glacial age of 12 000 years ago, with the surrounding swamp being formed later. Much of the importance of the area rests in the diverse range of habitats within close proximity to one another.

A crazy-quilt of habitats – wetlands and dry ground, open and closed, flat prairie and steep mountain areas – meet in a geological sweep ranging from 4,500 to 8,580 feet in elevation. At Pine Butte, the western edge of the High Plains grasslands edges up against cliffs and talus slopes, alpine meadows and montane forests.

> *The Nature Conservancy,*
> *Pine Butte publicity brochure*

The habitat comprises native foothills prairie, rocky ridges of pine and creeping juniper, spruce-fir forests, mountain streams, glacial ponds and spring-fed swamp. The reserve abuts the Bob Marshall wilderness area (federal/state reserve) and is an important extension of that protected area for migrating wildlife.

One of the area's most important inhabitants is the grizzly bear (*Ursus arctos horribilis*). Each spring, the grizzlies descend from their mountain retreat and follow the water courses down to the swamp to feed and raise their young. The bears quite easily replenish their energy reserves after winter hibernation in the rich wetland environment. Grizzly bears once roamed prairies, forests and foothills from the Pacific Coast east to Minnesota and south to Mexico, but retreated into a small portion of the northern Rockies when settlers moved westward. Today, some 800 of the original population of around 100 000 grizzlies remain. In 1975 they were designated a 'threatened species' under the Endangered Species Act 1973.

The grizzlies are often referred to as the 'indicator species' since their presence usually indicates wild, diverse, healthy environments, capable of sustaining a wealth of other animals, plants and natural communities. This has proved the case at Pine Butte Swamp, a peatland fen. Since taking over the area The Nature Conservancy has conducted regular inventories which indicate that the swamp houses a variety of plants, including the rare 'yellow lady's slipper', 'Macoun's gentian' and 'green keeled cotton grass'. To date, almost 40 distinct plant communities have been identified on the reserve. It is also the home of much fauna, including 43 species of mammals (such as muskrat, mink, moose, mountain lion, bobcat, lynx, black bear, mule deer, elk, beaver, coyotes and bighorn sheep) and 150 species of birds (warblers, waterfowl, waders and raptors).

The area is also rich in natural and cultural history. An area known as 'Egg Mountain' was unearthed by paleontologists in 1977. It comprises one of the richest finds of baby dinosaurs, hatched from the duck-billed *Maiasaura* ('good mother lizard') over 80 million years ago. Evidence suggests that the area around Pine Butte was settled as early as 8000 years ago. The Great North Trail, trod by Mongols who had migrated across the Bering Sea land bridge, cuts through the preserve. Tipi rings (stones used to fasten down the tipi), testify to the presence of prehistoric plains dwellers. Drive lines to a buffalo mire and a 'buffalo jump' (a steep ridge used to drive buffalo to their death before the horse was introduced to the USA) have also been uncovered. The area was later settled by European homesteaders and the Bellview Schoolhouse of 1906 is now used as the reserve's education center.

History of protection

Ownership of the preserve provides about as diverse a pattern as the natural habitats. Landowners around the reserve include the US government (Bureau of Land Management and Forest Service), the State of Montana and private landowners. The Nature Conservancy has easements with some of the private landowners to ensure that the management of their land is consistent with its own objectives. Part of the State land is managed for wildlife (mostly game species), the rest is leased by the Conservancy for conservation management. Some of the surrounding private lands with no conservation easements have been targeted for future purchase or easement negotiation by The Nature Conservancy. Limited resources force the Conservancy to seek partnership arrangements with landowners who might allow them to manage the land, without necessarily being subjected to the cost of purchasing full title. For example, a small site was targeted by the Conservancy for inclusion in the preserve. When it was purchased by a private landowner the Conservancy obtained a lease for management purposes and an easement was granted in order to protect it in perpetuity. Sometimes the advantages of not purchasing land outright are soon outweighed if the administrative costs of negotiating alternative protection are high. In this case the deal took three years to complete. With hindsight, the Pine Butte manager admits that purchase may have been a less costly option and certainly any future agreement of this nature would be conducted on the basis of the Nature Conservancy receiving an endowment for management.

Management

The greatest threat to the Pine Butte preceding the Conservancy's involvement came from the development of the area for residential or agricultural purposes. Unsuccessful attempts had already been made to drain the swamp (which had affected its natural balance) and the land around the swamp had been overgrazed. Despite very cold winters in this area there had been some residential development for summer homes and more was expected with Montana's rise in popularity as a recreational area.

The Conservancy ownership and easements over the area ensure that the swamp is now protected from development. Agricultural management of the area is vested in preserve managers, David Carr and Mary Sexton. An area which has been planted with nonnative grasses is mowed for hay by a local farmer who pays for 25% of the hay he takes, but provides fertilizer for the area as necessary. Native grassland around the swamp is leased in the summer and fall for cattle grazing (approximately 300 suckler cows and calves). In order to sustain growth, the grass must be grazed or burned. Local resistance to controlled burning has led to a grazing policy. Leasing the land for cattle also helps public relations with local farmers, who were concerned at first that the preserve would be 'locked away' for preservation.

The management team has looked into grazing bison on the preserve as an alternative to cattle. At present, bison meat affords a premium and their indigenous status would be consistent with the Conservancy's policy of protecting natural diversity. However, the bison, who are migratory, would need to be fenced onto the preserve and the type of fence required would preclude easy access for grizzly bears. The decision to retain cattle is typical of the type of compromises that must be made by individuals and organizations attempting to manage small areas of land for wildlife which, until very recently, enjoyed a very different land-use pattern.

Guest ranch

Located on the Pine Butte Preserve is the Pine Butte Guest Ranch, which was run as a private business from 1930 until it was purchased by the Nature Conservancy in 1978. The guest ranch is now operated by the

Fig. 6.3. Guest cabin at Pine Butte Preserve. Photograph: Victoria Edwards.

Conservancy and all proceeds from the ranch are used to support its conservation work. The ranch can accommodate up to 25 guests in eight cabins. These are made of stone and wood and are set amongst the aspen, cottonwood and fir trees that line the South Fork of the Teton River (Fig. 6.3). Each cabin has a large room with a fireplace, bedroom, bathroom and features handmade hardwood furniture. There is a central lodge and a Natural History Center, which is housed in an adjacent cabin, with reading materials, photographic exhibits, maps, fossils and a small selection of Nature Conservancy gifts. Guests also have the use of a heated outdoor swimming pool. Meals, which are served in the lodge dining room, emphasize homemade healthy food and include fresh produce.

The guest ranch combines normal relaxing activities with an in-depth natural history program. Two guided horseback trips are made each day to sites in the surrounding high country, and in the summer season daily treks are conducted by the ranch's full-time naturalist, focusing on plants, animals, geology and paleontology. Each spring and fall, outside the family holiday season, the preserve offers natural history workshops. A sample of the workshop program is shown in Fig. 6.4. Workshops are of four and seven days duration. The cost per person in 1994, which included food, lodging and transportation to the Preserve from a nearby airport, was $575 (4-day) and $1175 (7-day) (Fig. 6.5).

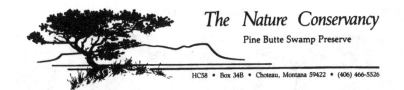

The Nature Conservancy
Pine Butte Swamp Preserve

HC58 • Box 34B • Choteau, Montana 59422 • (406) 466-5526

1994 TRIPS AND WORKSHOPS

RAPTORS OF THE NORTHERN ROCKIES
MAY 14–20

Join raptor expert Denver Holt for a fascinating week of study and observation of Montana's birds of prey. Daily field trips will lead to eagle nests, falcon roosts, nighttime owling, and more. $1175.00.

MONTANA GRIZZLY BEARS
MAY 21–27 AND SEPTEMBER 17–23

Search for the elusive grizzly with renowned bear biologist Dr. Charles Jonkel. Daily treks into occupied grizzly country, spectacular scenery, and stimulating audio/visual programs combine for an exciting week. $1175.00.

A NATURALIST'S TOUR OF THE ROCKIES
MAY 28–JUNE 3 AND SEPT 24–30

A splendid week of on-location learning in the wild Rockies and prairies of Montana. Enjoy many diverse topics and daily hikes into a wide variety of spring-graced (our Autumn guilded) landscapes. Led by Conservancy naturalist Ralph Waldt, with nearly 30 years' field experience in Montana, this workshop will focus on interpreting the land and its wildlife. $1175.00.

DINOSAUR DIG
SEPTEMBER 10–16

Don't miss this rare opportunity to be part of a real, working dinosaur dig site! One of the world's richest dinosaur sites is located on Conservancy lands just minutes from our guest ranch. The site is home to world-famous "Egg Mountain" and one of the largest dinosaur bone beds ever found. A paleontologist will guide us on-site as we explore for new finds and actually dig for bones. $1175.00.

NEW MINI-WORKSHOPS FOR 1994

NATURE PHOTOGRAPHY, October 3–7, 1994: Explore landscapes that are a photographer's dream glowing with rich autumn colors beneath the spectacular Rocky Mountain Front in the wilds of northern Montana. Plenty of "hands-on" learning on three daily field trips will combine with evening lectures and audio-visual programs to make this a fulfilling and memorable session. Taught by Jim Mepham, professional Montana photographer from East Glacier. $575.00.

A NATURALIST'S TOUR OF ROCKIES, Reading the Land, October 3–7, 1994: A brief but enjoyable version of week-long Naturalist's Tours described in this brochure. Three daily hikes will expose participants to a fascinating variety of landscapes in wilderness Montana. Lectures and slide programs are included. $575.00.

Fig. 6.4. Pine Butte Preserve and Guest Ranch program 1994. Source: The Nature Conservancy, Pine Butte Preserve.

Fig. 6.5. Guests on a 'mammal tracking' workshop at Pine Butte Preserve. Photograph: Victoria Edwards.

Outreach program

In addition to the fund raising recreational programs, Pine Butte Preserve has run an 'Educational Outreach Program' since 1990. The objective is to open up use of the preserve to local people and so encourage a greater enjoyment and understanding of natural areas. The donation of an old school house from a neighboring landowner helped to establish the outreach program. The Bellview Schoolhouse was moved onto the preserve and is now used as an education center. Most workshop and field trip activities occur near the school and a variety of educational materials is available for use by children and adults.

The program comprises educational presentations and field trips for local youth and school groups and summer workshops for children and adults (Fig. 6.6). A grant from Liz Claiborne (the owner of a successful American fashion house) supports the program, along with help from Conservancy volunteers and small participation fees. Details of the 1991 program are presented in Fig. 6.7.

The program has important public relations implications with the local community, who were very wary of the Nature Conservancy's presence when the preserve was first established. Involvement of the local residents in the Conservancy's work was seen as a major factor in contributing to

Fig. 6.6. Children from a nearby school hunt for fossils as part of Pine Butte Preserve Outreach Program. Photograph: Victoria Edwards.

Pine Butte's success. When the preserve was first established its objective as a refuge for grizzly bears was controversial, since many of the ranch holders in the area saw the grizzly as an irreconcilable enemy to livestock farming. Inviting local people onto the preserve and leasing part of the land for grazing has enabled people to observe that the grizzly and cattle may coexist on the land.

Potential costs of watchable wildlife programs

Nonconsumptive uses of wildlife might provide private landowners with an incentive to protect natural areas. The benefits of nonconsumptive use of wildlife can only be attained, however, at some cost. In particular, there are two general problems with allowing access to wildlife for recreational uses that private landowners must address. First, there are the financial costs of allowing participants of nonconsumptive wildlife use on private land. These may be either:

1. direct cost: the costs of providing specific services such as interpretation or the provision of visitor facilities; or
2. indirect costs: such as the restoration of damage to wildlife habitat and access routes.

In addition, the cost to the private landowner of disturbance to his or her exclusive right to enjoy the property should be taken into account. Careful planning and marketing is needed to enable landowners to achieve net benefits from opening their land to the public.

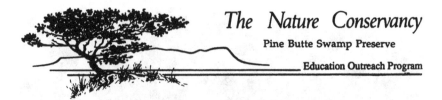

The Nature Conservancy

Pine Butte Swamp Preserve

Education Outreach Program

1991 SPRING AND SUMMER WORKSHOPS
BELLVIEW SCHOOL

May -   *Wildflower Workshop* for adults/children
        May 11; one day - crafts and identification; no fee
        Puppet Workshop for adults/children
        May 25; one day - make it and take; $5 for materials

June - *Wildflowers and Native Plants Workshop* for ages 7-14
        June 11-12; walks, crafts, identification, stories.
        Prehistory Workshop for ages 7 - adult
        June 13-14; geology, paleontology of Egg Mountain and
        Pine Butte with guest experts from Museum of the Rockies.
        Nature Writing Workshop for ages 7 - adult
        June 18-21; creative writing with poetry, fiction, non-fiction
        lead by Ripley Schemm Hugo and Mike Riley.

July - *Introduction to Pine Butte Workshop* for ages 7-14
        July 15-18; Plants, Animals, Dinosaurs, and History
        (For first-time participants in Pine Butte workshops)
        Blackfeet and Buffalo Workshop for ages 7-18
        July 23-24; Archaeology and History with Bud Olson.

August - *Introduction to Pine Butte Workshop* for ages 7-14
        August 5-8, (see above)
        Mammal Tracking at Pine Butte Workshop for ages 7-18
        Aug. 19-22; hike, explore habitat, cast tracks with experts.

Instructors: Mary Sexton, Debbie Carlson Saylor (certified teachers), Ralph
              Waldt (Preserve Naturalist), and guest experts.

Costs: $10.00 long course/ $5.00 short course for local and regional
         participants. $25.00 long course/ $15.00 short course for others.
         A grant helps defray costs for local participants.

Enrollment: 20 participants per course limit.

Contact: Mary Sexton   Phone: (406) 466-5526
            Pine Butte Swamp Preserve
            HC 58 Box 34B
            Choteau, MT 59422

Fig. 6.7. Pine Butte Preserve Outreach Program 1991. Source: The Nature
Conservancy, Pine Butte Preserve.

Damage to habitat and wildlife

A second, and arguably more alarming problem, is the irreversible damage
that visitors might cause to the wildlife and its habitat. Despite the term
'nonconsumptive' use of wildlife (meaning that there is no intended removal
of wildlife or wildlife habitat in any of the listed activities), nonconsumptive
pursuits may have an affect on wildlife resources. Indeed, some authors (such
as Wilkes, 1977; Weeden, 1976) reject the concept that some outdoor recreation activities are 'nonconsumptive' and note the impact of such activities

on vegetation, wildlife and the quality of the environment. White & Bratton (1980) claimed that direct and indirect human disturbances were the greatest threats to protected areas and argued for the need to manage and protect reserves from overuse, while Edington & Edington (1986) stated that 'evidence is now accumulating to show that animal watchers can represent a serious source of disturbance even in protected areas.'

Several texts explain the type of damage that can be executed by 'nonconsumptive' wildlife use (see, for example, Goldsmith, 1974; Satchell, 1976; Schoenfeld & Hendee, 1978; Edington & Edington, 1986). Problems associated with visitors stem not only from the disturbance caused by an area accommodating a large number of people at any one time, but also from the behavior of the visitors. For example, the problems associated with the feeding of wild animals by visitors and subsequent changes in the social behavior of the animals is well documented throughout the world. Glick (1991), commented that the feeding of coyotes at Yellowstone National Park meant that 'they have now replaced bears as Yellowstone's most common roadside beggars.' He reported that the effects of feeding had resulted in 'several episodes of emboldened coyotes threatening park visitors' and the offending coyotes being shot by Park Service officers. Feeding by visitors can make also species more vulnerable to other species: this is particularly true of nesting birds, which may be encouraged to leave their nests by offers of food, leaving their eggs open to predation from other species. Disturbance by visitors can also separate breeding pairs, resulting in a subsequent fall in reproduction of the species.

Sadly, not all of the problems encountered with watchable wildlife can be attributed to the behavior of visitors. Over anxious reserve managers might also interfere with feeding of wildlife species in order to encourage the animal's presence. Recognizable effects include changes in diversity or abundance of wildlife at a site, or an altered habitat structure and utilization. However, such behavior should not be associated necessarily with *private* reserve managers and owners. Interference with wildlife has been taken to an alarming extreme at Gir National Park in India, where wardens tether buffaloes so that visitors may obtain a better view of the lions attacking them. The following extract from Insight Guides, *Indian Wildlife* (Israel & Sinclair, 1987) on Gir National Park and Sanctuary illustrates the pitiful state of some watchable wildlife programs:

The best way to see lions is, of course, in their natural surroundings at dawn and dusk, when these predators are on the move. This can be done from a car as, owing to the protection given to them, they are not shy of motor vehicles. Nevertheless, one cannot be certain about seeing them on all occasions. The Forest Department arranges lion shows every Sunday which become a *mela* of scores of people watching lions attracted by a buffalo.

Numerous accounts of the incompatibility of visitors and wildlife might leave readers with a pessimistic outlook on the opportunities for protecting natural areas through recreational provision. Some authors place the import-ance of visitor safety in National Parks above the existence of wildlife. One author has called for the removal of grizzly bears from certain parks, to make the parks more suitable for hiking and camping (Moment, 1970). Indeed, in Yellowstone National Park, conflicts between visitors and bears at Fishing Bridge Campground and Visitor Center have reportedly resulted in the death or removal of more grizzlies than at any other site in the Greater Yellowstone Ecosystem (Glick, 1991). Other authors have listed the appalling damage to habitats caused by public access, for example Willard & Marr (1970, 1971) reported the effect of easing public access to alpine meadows in the Rocky Mountain National Park, afforded by the construction of a high level road, that resulted in destruction of the vegetation cover of up to 95% in areas close to the road.

The management of watchable wildlife programs on private land, however, need not reflect such problems of combining visitors and wildlife. Indeed, private enterprises are equipped to overcome such problems by virtue of their ability to restrict access and to monitor and control the behavior of their visitors. Aside from the landowner's ability to derive income by charging for access, other benefits stem from the exclusion of the general public from the land. First, most nonconsumptive activities are rival: that is, the extent of one person's enjoyment is affected by the number of other people allowed access to the wildlife area. This is particularly true of observing and photo-graphing wildlife, where the amount of human presence may be directly pro-portional to the inclination of the wildlife to remain in an area. Exclusion allows the landowner to carefully monitor participants' enjoyment and arrive at an optimal number of participants' at any one time. Second, exclusion allows the landowner to restrict access over time and space. Edington & Edington (1986) referred to the problems encountered with bird watching in Britain when there was open public access to a site:

Reports of a rare species are liable to produce an influx of hundreds or even thousands of observers to a site, each intent on seeing the bird in question and if possible photographing it at close quarters. Frequently on these occasions little attention is given to the welfare of the animal itself or the damage which might be caused to the habitat (*British Birds*, 1982).

Several authors have cited the importance of controlling visitor access to sensitive sites. Faro & Eide (1974) reported on how a limited permit system was necessary at the McNeil River State Game Sanctuary to reduce the undesirable effects of public overuse of an area for viewing brown bears. The control afforded by private land may allow a flexible management practice to evolve, which restricts access to participants in areas sensitive to disturbance, either exclusively or at certain periods of the year. In Montana, Pine Butte Preserve managers prepare an annual 'travel plan' which sets out the type of access that will be allowed on the preserve for the forthcoming year. The plan restricts access over all of the preserve to **all** persons, including Conservancy employees. It is an attempt to reconcile human use of the area (which is good for fund raising and goodwill) with the demands of sensitive wildlife: particularly the grizzly bear, a species most susceptible to disruption by human activity.

Fee-paying watchable wildlife enterprises may hold an advantage over free public lands when monitoring and controlling visitor presence. The presence of an excess of visitors at a site, and consequential degradation of the habitat and wildlife, may take a while to reveal itself to the managing authority. Visual evidence of environmental damage may be the first indication that visitor numbers have increased to the extent that they are jeopardizing the existence of the very features being visited. In such cases, habitat and wildlife restoration programs may have to be introduced along with new measures to control visitor numbers. Lindberg (1990), in a survey of visitors to the Spanish Peaks Primitive Area in the USA, found that when the number of trail encounters with other visitors increased to a certain point (in this case, five encounters), the willingness of visitors to pay for access to an area can decrease rapidly. An enterprise that is charging a fee for access should receive very clear signals from its customers when the 'human carrying capacity' in terms of visitor enjoyment has been reached. This might well provide a crude but early warning that it is time to review visitor numbers.

Conclusion

An estimated three quarters of all Americans participate in natural resource-based recreation and most of this recreation is heavily dependent on public land. Increasing pressure on public land, accompanied by overcrowding and a deterioration in the user's experience, suggests that private landowners might start to provide areas for recreational use. Many areas of private farm and forest land harbor an impressive variety of plants and animals that might be managed for nonconsumptive use. Aside from the more traditional production of agricultural goods, the resources might be managed in a way that will attract an affluent sector of the American population, who merely want to enjoy nature. Indeed, there is an increasing demand from higher income earners for access to natural areas. Cutler (1989) remarked that: 'To ignore this opportunity and plow down or cut down this habitat may well be to kill the goose that laid the golden egg.'

Managing natural areas for consumptive and nonconsumptive recreational enterprises is not a panacea for *all* conservation problems: some areas simply cannot sustain such uses. In all areas there are potential problems that must be addressed if management is to maximize the environmental benefits of each enterprise (see Fig. 6.8). Careful planning for visitors to natural areas can help managers to maximize the economic benefits of visitor enterprises, while minimizing environmental costs. Figure 6.9 provides a checklist for managers planning a recreational conservation enterprise.

While attempting to mitigate environmental costs, landowners and managers may find that they also face political opposition to their proposals. The management of private land for free public access, particularly for hunting and fishing, has a strong tradition in the United States and landowners establishing fee-paying enterprises will meet with some fierce opposition in the form of public management advocates. Arguing that fee-hunting provides opportunities for public consumption of wildlife for all but the privileged classes, the public management philosophy rejects the possibility that private management of natural areas might be capable of combining sound conservation practice with a financial return on capital. Certainly there have been poor examples of private wildlife management, especially hunting enterprises: some game ranches, with their introduction of exotic species and manipulation of herd development for trophy improvements have provided fair ammunition for opponents of private resource management. However, such poor examples are not exclusive to the private realm of natural resource management: for example, Chase (1987) provided a comprehensive account of management failures in Yellowstone National Park; and Dowdle (1981) and Deacon & Johnson (1985) were similarly critical of US public forestry management.

To focus solely on the potential costs of private resource management is to ignore the scope for reaping huge public benefits. While land remains in the private domain, landowners will continue to seek ways of securing an income flow that supports the capital value of their asset, the land. Alternative uses for land, such as intensive agriculture and forestry, subdivision, mining and quarrying, will compete with the possibility of managing land in its natural state in the decision-making process of each landowner. Deriving income from fee-paying enterprises that are capable of returning revenue from the

ADVANTAGES

incentives
1. Provides incentives for landowners to enhance management of the land for wildlife
2. Provides incentives to maintain public access
3. Reduces the comparative benefits of subdivision, intensive cropping, mining and quarrying
4. Produces income to offset costs of wildlife damage to cultivated forage areas

control
5. Enables landowners to control numbers of visitors and so minimize their impact on environment and property
6. Enables landowners to control behavior of visitors and exclude visitors from particularly sensitive areas
7. Enables landowners to control visitor behavior in order to enhance the visitor experience and safety aspects of the pursuit

information
8. Provides information on changing demands of visitors and expectations of 'wildlife experience'
9. Provides landowners with the opportunity of gaining a greater understanding of natural area management and habitat restoration

DISADVANTAGES

perverse incentives
1. May provide incentives for herd manipulation for trophy stock or wildlife interference for viewing
2. May encourage the importation of exotic species that can damage the natural environment and native species

ideology
3. May detract from visitor enjoyment by introducing the concept of paying for a wildlife experience
4. May be politically and ethically unacceptable to advocates of public use and management of natural resources

costs
5. The increase in wildlife on the land may result in crop damage to adjacent land

Fig. 6.8. Potential advantages and disadvantages of fee-paying recreational enterprises.

DO

1. Carry out an environmental inventory on the site to establish a benchmark
2. Set clearly defined objectives (environmental, social, cultural & economic) for the land-owner(s), staff and visitors
3. Make an assessment of the likely impact of the enterprise and establish a tolerable level of visitors
4. Minimize any environmental damage antici-pated from the proposed enterprise
5. Know your market and develop services accordingly
6. Employ local people and contractors where possible
7. Develop working rules for staff and visitors
8. Monitor visitor satisfaction and adapt the oper-ation of the enterprise accordingly
9. Continuously review the effectiveness of the interpretive material
10. Continuously monitor the effect of visitors on the environment

DON'T

1. Be tempted to overexploit the site by increasing visitor numbers beyond its capacity – remember the site is your goose which lays the golden egg!
2. Manipulate wildlife behavior or numbers for the visitors' sake: use imaginative interpretation of what you do have instead

Fig. 6.9. Ten 'Dos' and two 'Don'ts' for establishing a conservation recreational enterprise.

land in its current or improved natural state provides a sensible solution to both the problems of increasing public demand for recreation and the need to manage land as a sustainable, yet productive, resource.

7

Turning development into conservation

Land-use ethics are still governed wholly by economic
self-interest, just as social ethics were a century ago.
Aldo Leopold (The Land Ethic)

Introduction

Chapters 5 and 6 examined how landowners might derive additional income
from natural areas by charging for public access and use of their land. By
harnessing the private 'use' benefits to be gained from natural areas, a whole
host of fee-paying enterprises can create a market for conservation without
the need for the land ownership to change hands. A more sophisticated range
of property rights is developed, which extends transfer beyond that of freehold fee-simple. Both chapters illustrated how additional income, derived
from such practice, acts as an important incentive in encouraging landowners
to protect natural habitat.

This chapter examines how the use and nonuse benefits of owning and
living in a conservation area might be exploited to facilitate the protection
of natural areas. Rather than providing an additional *income* source from the
natural area, the benefits of conservation are *capitalized* into the freehold
value of a site. Careful use of protection mechanisms ensures that the area
is legally protected in perpetuity. In practice the powerful incentive of private
landownership, often coupled with the ability to reside in the area, ensures
that the protection of the natural area is continuously enforced.

Conservation real estate

In developed countries, one of the greatest threats to natural areas and wildlife
management must be that posed by the construction of buildings, roads and
engineering works. Many landowners, who for years were content to derive
an annual income from their land, are persuaded to sell their estate for subdivision when a large capital sum is required. The need to realize the *capital*
value of the land might arise when children are ready to attend college or
the landowner wishes to retire from farming the land or a demand for estate

or gift tax is received on transfer of the property. Land that has been subdivided may have a reduced chance of being managed for nature conservation purposes. Fragmentation of the ownership could physically affect the land's ability to sustain species. In addition, it can be argued that the family ownership and management of an area, going back several generations, may be more sympathetic to conserving the natural features of the land because of the connection with such features over a long period of time. New owners of land, without such emotional attachment, could be more minded to extract the best financial return possible from their land, having less regard for the manner in which it is managed and the conservation of plant and animal life over time.

Until recently, conventional thinking has been that conservation easements and other restrictions on property pose obstacles to the successful marketing of that property. In some cases, however, they may be marketing tools. When landowners live in an area of exceptional beauty and diverse wildlife habitat, they are often keen to ensure that it is protected, not only during their lifetime, but for the benefit of their successors in title. Conservation real estate agents have acknowledged such feelings by developing a new market in America. Conservation real estate is simply a process by which property with a high conservation value is sold to purchasers who agree to preserve the area and keep it undeveloped. A conservation easement is placed on the property, to ensure perpetual protection of the natural area.

Marketing property in this way is often employed by land trusts who wish to realize the capital value of a site they own without threatening its continued protection. Originally, the site may have been purchased by the trust or donated to it by a benefactor. Placing an easement on the property and selling it on enables the trust to use the freed-up capital on a subsequent project. In addition, it lessens the future management costs of the site by passing on the responsibility for management to the new owner. The strength of the deal lies in finding a suitable 'conservation purchaser'. If this can be achieved, then the future land use and management of the site should be compatible with the trust's objectives and the easement will not be violated.

Conservation brokers

A conservation real estate broker is normally used to oversee the transaction. This has several advantages. First, the broker has the knowledge and experience to ensure that the transaction progresses smoothly: very few conservation organizations have such experience. Second, the broker is only paid if a successful deal is completed. If the conservation organization were to attempt to carry out the preparations and negotiations, then they might expend a lot of time and resources on an unsuccessful transaction. Third, it is possible that a conservation organization will lose the trust of its members if it earns the reputation of selling land on to individual wealthy purchasers. Many members will see it as encouraging 'gentrification' of the countryside and could well resent an organization that is seen to be operating in the property market.

Seeing an opportunity in the market, several brokers now specialize in selling conservation properties to conservation buyers. Rather than waiting for a conservation organization to present them with an area of land for sale, the conservation realtors will attempt to find conservation buyers for any property that has a high natural area value. The specialization has enabled them to earn the trust of local conservation organizations and buyers alike. Often they will be the first to learn of any land that is to be put up for sale. If the property is threatened by development and has a high conservation value, then the broker may learn to approach the conservation organization in the first instance. The conservation organization will then help the realtor to find a purchaser or group of purchasers who will protect the area, normally by placing advertisements in the organizations's newsletter.

Benefits

There are obvious advantages to this procedure for the purchaser of a piece of conservation land who wishes to keep the land undeveloped. The purchaser, by virtue of the tax deductible nature of the easement, is able to purchase an area of land but claim back part of the purchase price in the form of taxable deductions. Such a practice can significantly reduce the net purchase price of the land. The conservation organization also benefits and not only from the ability to add another easement to its portfolio. It may also receive a donation from the broker, if it has been instrumental in finding a purchaser.

Conservation real estate is a practical means of ensuring that an area of land is protected, while minimizing the costs to any one party. It is a fine example of how the conservation movement has refined the market for conservation by seizing opportunities in the property market to achieve its own objectives. It is important to appreciate, however, that the success of the transaction often relies upon the tax deductible easements. Paul Brunner (1982) cited three incentives for conservation buyers:

1. a future investment – something to hand on to future generations
2. hunting and fishing properties
3. tax shelters.

Case study: American Conservation Real Estate

'American Conservation Real Estate' (ACRE), is a licensed real estate broker and consultant based in Helena, Montana. The Company acts as a buyer brokerage, offering research and consultancy services to people who wish to purchase land in Montana, Wyoming and Idaho. The company ensures that it understands the requirements of its purchasers well and locates available properties that will serve their needs. By establishing a consultant relationship with its clients, ACRE is able to counsel them on land use issues and the social and cultural aspects of living in isolated rural communities. In addition, it sells land on behalf of conservation minded landowners. The vendors of conservation properties who wish to see their land respected and protected by future landowners may place an easement on the land (if no such restriction already exists) before sale. The company is clear that the primary objective of its activity is conservation:

We discourage rural land subdivision and encourage long-term conservation of open space, wildlife and fisheries habitat. We advocate land use that is consistent with the region's climate and rainfall. We encourage the maintenance of historic and family-owned farming and ranching businesses. We encourage conservation through the donation of conservation easements and thoughtful stewardship and management. We give financial support to local land trust organizations.

Coulston, personal communication 1994.

ACRE has been involved in several interesting deals since its establishment. In some cases, clients are able to provide clear specifications of their requirements. In one case study, a client was specifically interested in purchasing large tracts of land where riparian corridors and wetlands could be protected and restored to provide a wide range of wildlife habitats. Over a period of two years, two traditional ranches of approximately 2000 acres were identified and purchased. The first ranch comprised native grassland and timber with several miles of river corridor. A conservation easement, which will limit future development on the ranch and protect native grasslands and the river corridor, is currently being prepared with the Nature Conservancy. In the meantime, the river corridor has been fenced to keep livestock away from stream banks and the margins of the river. The second ranch was purchased to remove intensive livestock grazing in order to restore large areas of wetlands adjacent to the river corridor. The US Fish and Wildlife Service has met with the new owners to discuss restoration of waterfowl nesting and feeding habitat. Negotiations are taking place with the Nature Conservancy and the Fish, Wildlife and Parks Department concerning an easement over the land. Eventually livestock may be returned to the area for grazing in controlled numbers and thus used as a management tool.

ACRE has tried to shake off the popular image of 'the West' being bought up by celebrities eager to escape the pressures of Hollywood life. Their client list includes families from New York, Florida, Nebraska, Washington and New Hampshire, looking to expand their lives in the west and who are attracted by the prospect of becoming members of a ranching community, fishing in scenic rivers or enjoying the outdoor life. Coulston (personal communication 1994) claimed that such people have a genuine interest in enhancing and protecting natural areas while wanting to be regarded as long-term members of the community, he confirmed that

American Conservation Real Estate Company has found that conservation buyers regard the easement as an asset to the property. Buyers are betting on long-term preservation of habitat as it disappears elsewhere.... Conservation easements reassure buyers that potential problems have been reviewed by people with expertise. Stewardship plans on lands with conservation easements are like having free consultants about water, timber management, and neighboring land uses for the new owners of the property.

ACRE also lists properties for vendors who seek conservation buyers. In some cases, where no conservation easement already exists one is placed on the property prior to sale. In other cases, a conservation easement might be specified as a condition of sale. American Conservation Real Estate Company has recently marketed a 1960 acre ranch at the edge of Yellowstone Park that has had a conservation easement in place for over ten years. The easement provides for open space to honor exceptional wildlife habitat for bears, mountain sheep, elk and moose. In such cases, prospective buyers must recognize the importance of the private property as, effectively, a preserve.

Case study: Conservation Solutions

Conservation Solutions is a conservation consultancy located in Sacramento, California. The overall goal of the company is 'to develop new sources of private funding for land-based conservation programs, while simultaneously providing a client responsive land search and brokerage service.' (Werschkull, personal communication 1994). The company achieves its aim by acting as an environmental consultant, providing advice on matters such as sustainable project development and environmental audits. One of the several consultancy services offered by the company is a land search and brokerage service for conservation buyers. The company is not licensed as a real estate broker, but works through selected, cooperating real estate agencies. The service offered by Conservation Solutions usually begins with the prospective purchaser completing a questionnaire. A copy of the type of questionnaire sent to clients is illustrated in Fig. 7.1. On completion of the questionnaire the client meets with a member of Conservation Solutions, which acts as both an information source of potentially suitable properties and a facilitator for discussion among family members and other individuals who have an interest in the acquisition. Once the company has defined the client's requirements, it contacts various resource agencies, organizations and realtors about potential properties. Detailed information is provided on potential properties, including environmental assessments with analysis of the site's constraints. In addition, site visits are organized for the client group at this stage. Once a client decides on a property, Conservation Solutions locates a broker or attorney to represent the client. During negotiations and purchase, the company remains available for any nonbroker related tasks. On completion of a successful sale, Conservation Solutions is paid a 3% 'finders fee'. Should no sale take place, then hourly fees and costs are paid.

A hypothetical transaction provided by Conservation Solutions involved the purchase of a 250 acre property along the Mokelumne River, California. The search for the property began by convening a meeting with the family and the development of a property search criteria. The family specified the elevation that they did not want to be above, the cash and long-term capital they were willing to commit to a purchase and the absolute requirement that river frontage be included. The family members appreciated that the optimum conservation purchase would not include a road or structure in proximity to the river: indeed, they actually preferred the idea of walking to the river from a homesite located outside the immediate river corridor. Conservation Solutions contacted friends in the Amador, American River and Placer Land Trusts. After approximately two months of telephone calls, the company assembled a list of four potential properties for site visits. One family member, a college-student interested in environmental planning, visited the properties with the company on behalf of the family. Photographs were taken and a summary report was presented to the family, highlighting key property characteristics. A family site visit to two of the properties resolved the debate over purchase in favor of the property on the Mokelumne River, which has impressive stands of black oak, buckeye and digger pine. The sale was completed in two months with the assistance of a broker affiliated with the Amador Land Trust. On completion of the sale, a conservation easement over the 200 acres along the Mokelumne River was granted to the Amador Land Trust. The family retains full title to the remaining 50 acres, on which a home will be built. The site provides the most fire-secure location on the property

CONSERVATION
SOLUTIONS

Conservation Land Search Questionnaire

The Conservation Land Search Information' Sheet and Retainer Agreement provide information about our Land Search Services. You may find it helpful to review these documents as part of completing this Questionnaire.

Key Information

1. What is the approximate dollar amount you are intending to commit to a purchase?

2. When could you be prepared to make this purchase?

3. Are there special circumstances that encourage purchase by a specific date (such as capital gains taxes on the sale of another property or business)?

4. What geographic area or areas are you interested in?

5. Are you open to a variety of acquisition and protection opportunities or are there particular types of habitats and biological resources that you are most inspired about protecting through your purchase?

Acquisition Objectives

1. Please list your objectives for the property purchase Conservation Solutions will be assisting you with.

2. What would be your schedule (if you have one) for any capital improvements to the property once it is purchased?

3. Is it necessary for this property to have income-producing capability?

Your Conservation Interests and Experience

1. Are you familiar with the concept of a conservation easement and how nonprofit land trusts operate? Would you like more information on these subjects?

2. What are the experiences that have shaped your perspectives concerning conservation? Do you feel a particular affinity for the mission of a certain conservation organization?

3. Would you like land trust and conservation/land protection information for a particular geographic area?

4. Are you familiar with the public and private programs now underway to restore native habitat? Would you like information on these programs?

Other Information

1. Please list any other information or issues that you think could be relevant to your evaluation and selection of a property with conservation significance.

2. Are there other family members that may participate in evaluation and selection of a conservation property? Would you like us to send them this Questionnaire?

3. Please list the addresses and telephone numbers of locations where you prefer to be contacted as we proceed with your Conservation Land Search.

If you have questions or would like to discuss any of the above, please call Grant Werschkull at (916) 325-4800 (office) or 455-2923 (home).

Thank you for your interest and commitment to private-sector conservation. We look forward to working with you!

Fig. 7.1. Conservation Solutions: Purchaser's Questionnaire.

and ensures that the preferred conservation site can be maintained and enhanced without conflict from development.

Further developments – securing deals on threatened land

Lands are often offered for sale when the owner experiences financial distress. In such cases, a rapid sale may be necessary in order to mitigate any financial problems. Unfortunately, a rapid sale may not result in the best conservation deal. While conservation real estate brokers are experienced in marketing and attracting conservation buyers, they often need more time to prepare the property information and work through networks to match the property with the right buyers for long-term stewardship commitment. In such cases, a temporary and sympathetic purchaser would provide the opportunity to plan a future conservation sale. Such purchasers would need to be prepared to take a financial interest in the property and control its management until a long-term conservation buyer can be found. Lane Coulston of American Conservation Real Estate is working on such a solution: 'I am confident that with a little more time and strategic marketing, we can influence the outcome for the highest and private conservation solution.' (Coulston, personal communication 1994).

Partial development

'Partial Development' is an increasingly popular form of natural area protection in the USA. It refers to a process whereby a conservation organization purchases a property and then funds its acquisition by allowing development on part of the property, but ensuring that critical areas are preserved. Essentially, it is a private mechanism for controlling the amount and type of development that can take place in a given area. It allows the objective of protecting natural habitat to be achieved with little cost to the organization. Indeed, in some cases enough revenue is made from the development deal to fund a subsequent conservation project.

Partial development of an area is often initiated by a conservation organization. When an area has been offered for sale to a developer, and the organization has identified that it is a site with some conservation value, they will attempt to put together a deal that will allow development of the site but in a restricted manner. The procedure for assembling a partial development deal is long and may be expensive. First, the site is identified and a complete resource inventory is conducted, including the site's agricultural capability, its wildlife habitat and any development limitations. The inventory entails a full assessment of the physical characteristics of the site, including soil, topography, climate, habitat, etc. Second, any development potential is identified, an appropriate area is selected for partial development and a development site appraisal is completed, specifying the type and density of development. Third, the price to be charged for the development site is calculated. The most frequently used method of valuation is to calculate the *cost* of completing the development and deduct it from the *price* that might be achieved for the properties once completed. Taking into account a suitable percentage for the developer's profit, the appraiser may then arrive at a figure that represents the value of the site to a developer.

Once the price of the development site has been determined, a developer is sought who will work with the conservation organization to achieve a suitable end product. In the selection of a developer, emphasis is placed on the compatibility of the developer with the organization and not just on the ability to pay the determined price. Once bids have been solicited from developers, interviews are conducted and the developer's past history investigated. If a suitable developer can be found, then the organization will design a plan for the land, which will ensure the perpetual protection of the specified natural areas. It is only after all these tasks have been successfully completed that the organization will submit a bid for the land offered for sale. If the organization is successful in bidding for the land, then it is important that the two deals (the deal it has set up with the developer and the deal to purchase the land) are completed simultaneously. A step-by step guide to the development process is presented in Fig. 7.2.

Costs

Both the Nature Conservancy and a growing number of local land trusts are experimenting in the negotiation of partial development transactions. In some instances, the trusts have been given easements as a condition of the developer's land-use permit. In others, property has been purchased by or donated to the trust and has been resold with restrictions on its development.

There are some problems in entering into partial development agreements for the conservation organizations concerned. First, there is the possibility that the organization may loose the support of its members if they regard the agreements as a compromise to the organization's conservation commitment. Second, should an organization enter into partial development deals on a regular basis, there is a danger that the Internal Revenue Service (IRS) may regard the activity as 'for-profit' transactions. Ultimately, the conservation

Step One

- Identify site
- Conduct resource inventory
- Evaluate site for:
  agricultural capacity
  wildlife habitat
  development limitations

Step Two

- Identify development potential
- Select appropriate area for partial development
- Specify type, design and density of development

Step Three

- Conduct project appraisal:
  Marketable price of completed properties
  LESS cost of completing development
  <u>LESS developer's profit and finance charges</u>
  =   Value of site to the developer

Step Four

- Identify developer who:
  can pay the price
  is compatible with the conservation
  organization's objectives

Step Five

- Design future management plan for site

Step Six

- Purchase site   } Simultaneously
- Sell site on to developer

Fig. 7.2. Step-by-step procedure for partial development.

organization could lose its nonprofit making status and be required to pay income tax on any transactions completed. Third, when a natural area adjoins a private development, it may be difficult to combine protection of the area with public access. This may be politically detrimental to conservation organizations, which rely on allowing access to their preserves in order to increase support for their objectives.

In addition, partial development is not appropriate for the protection of all types of natural areas. First, the individual or group initiating the scheme must be satisfied that the area of land set aside for development is not critical to the survival of the whole habitat. Second, a conservation-minded purchaser must be found who will protect the natural area and abide by the conservation easement. Third, there must be a market for the type of development proposed.

In some cases, the partial development could incorporate a number of additional features to ensure natural area protection. For example, a homeowners' association might be established for the residents of the development. This might allow the conservation organization to part with all of the land, placing an easement on the undeveloped land and passing it on to the homeowners to be held in common. The immediate management of the land may then be conducted by the residents. For example, if some of the land is suitable for grazing, the homeowners' association could enter into a grazing lease, earning some income for themselves and relieving the conservation organization of future management costs.

DIY development

It is possible for a conservation organization to conduct a development itself, without the need for entering into an agreement with a developer. Such practice involves higher risks for the organization, but should result in a higher financial return. However, there remains something of a dilemma for any organization wishing to conduct development in this way. While the organization would probably need to gain considerable experience in this field to be successful, the Internal Revenue Service is likely to withdraw the tax-exempt status of an organization engaged in a large amount of development work. The ironic partnership between developers and conservation organizations therefore seems likely to continue.

Recognition that natural areas might contribute to the capital value of nearby freehold property has stimulated other groups of individuals to initiate the protection of natural areas. In its 1985 survey of conservation easements, the Land Trust Alliance found that 29% of all dealings reported were with 'for-profit landowning entities: developers, land investors, or for-profit corporations' (Land Trust Alliance, 1985). Conservation real estate is becoming big business in the USA: particularly in rural areas traditionally important for farming, but which are now being colonized by migrating urban dwellers. It is likely that having adjacent preserve land will prove to be an added attraction when marketing residential property. If the land is retained by a conservation organization, then access permission for neighboring residents may be necessary.

In some cases, allowing partial development of an area may well result in more harm than good, by introducing development pressures and residential conflicts into relatively undeveloped natural areas. However, where development cannot be avoided, partial or limited development may be the only way to protect natural areas.

Case study: The Virginia Coast Reserve

The Virginia Eastern Shore is a 70 mile long peninsula, bounded by Chesapeake Bay to the west and the Atlantic Ocean to the east (Fig. 7.3). The islands hold great ecological significance: they contain the largest colonies of nesting shorebirds on the east coast of the USA; their marshes support major concentrations of migratory waterfowl; in the spring, bottlenose dolphins migrate along the Atlantic coast, just off the islands; and in summer, juvenile loggerhead sea turtles inhabit the shallow coastal bays.

The Nature Conservancy has been working to protect areas of the Virginia Eastern Shore since 1969. It has acquired all or part of the 14 barrier islands and so created a 35 000 acre reserve, starting 55 miles from the Maryland/Virginia border and continuing to the tip of the peninsula. Since acquisition, the barrier islands have been designated as a National Natural Landmark. However, the Conservancy has recognized the importance of areas around the islands as part of the protective system. The coastal bays between the islands and the marshlands to the west provide rich spawning and nursery ground for fish and shellfish. The nesting birds and wildlife depend upon the food found in these bays and along the mainland forests and creeks of adjacent agricultural land. The mainland forests and fields also provide a migratory corridor for raptors, songbirds and other types of upland birds.

Recognition of the mainland's importance led the Conservancy to identify that 'conserving the wildlife resources of the Eastern Shore calls for owners of these properties to continue to respect the Shore's traditional values for land and wildlife.' (The Nature Conservancy, 1991). The Conservancy needed to move away from its traditional approach of protecting rare species and to consider how that relates to large, regional areas. In doing so, it began to recognize ecosystem level patterns and processes over the large area of the barrier islands, coastal bays and creeks, salt marshes and mainland that make up the Eastern Shore. It identified that the long-term protection of the barrier islands could not be separated from long-term protection of the surrounding ecosystems. In effect, it had to manage the 'greater coastal ecosystem' in its entirety.

While inappropriate agricultural practices and inputs threatened the preservation of the reserve, perhaps the greatest threats came from development. The urban Chesapeake Bay region is growing by more than 85 000 people

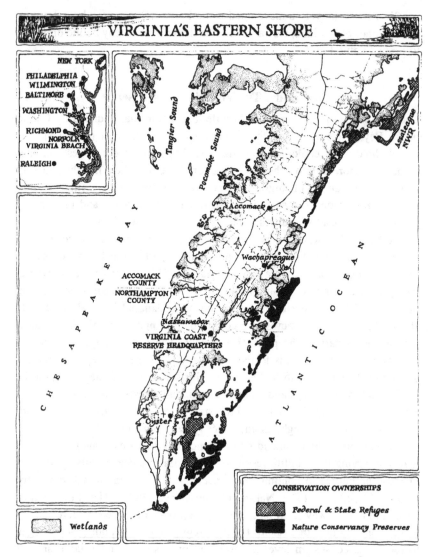

VIRGINIA'S EASTERN SHORE

Fig. 7.3. Virginia's Eastern Shore – Protected areas. Source: The Nature Conservancy, Arlington, Virginia.

each year. 'Condominium and commercial development has exploded across Virginia Beach's bayfront and oceanfront. A more scattered, sprawling sort of development is pushing south along the Atlantic coast of Maryland.' (*Newport News*, 1986).

Conservation plans

As a solution to agricultural and development threats, the Conservancy not only began to work with local landowners designing conservation plans for the farms but also acquired several of the farms. The intention was to resell these properties to buyers who would become 'conservation partners': private landowners who will legally agree to follow a conservation plan while enjoying ownership of the land.

Properties have been purchased from the Maryland border to Cape Charles at the tip of the peninsula. The mainland part of the preserve currently comprises an additional 8000 acres of protected land. Through the project, the Conservancy hopes to set an example of responsible, sensitive development. Several farms have already been resold in this manner. Conservation easements are attached to the land's title deeds, which will:

limit the property to low-density development;
establish 'buffer' strips to protect particularly sensitive areas;
enforce ecologically sound agricultural practices;
preserve the wetlands and waterways; and
restrict commercial logging or clear cutting in the woodlands.

A conservation plan is drawn up for each site and presented to potential buyers. Buyers are sought who understand the importance and value of the restrictions. The land has been sold in various parcels ranging from 35 to 800 acres. The family who owned one of the smaller farms recently sold to the Nature Conservancy cited two main reasons for sale:

1. the poor value of the farm for agricultural use terms of its size and productive capacity; and
2. the real estate taxes on the land, which were high because of its development potential.

'Escalating property values and real-estate taxes are seducing farmers into selling waterfront land that has been in their families since it was granted to them by the King of England.' (Badger, 1990). The rental income from the farm did not cover the Virginia property taxes.

Since the family did not want to see the farm sold for high density development, a partnership was formed with The Nature Conservancy. The Conservancy conducted an environmental assessment of the property and then searched for a buyer. A conservation easement was drafted, a conservation marketing package was designed and within a few months two buyers were found. Each buyer has a small tract of land where a house may be built: the remaining 150 acres will be held as a saltmarsh and woodland preserve.

Taxes – getting the incentives right

This chapter has explained how the formal protection of natural areas might be regarded as an asset rather than a liability in the freehold title to the land. It suggests that conservation and residential development of natural areas can be compatible with the protection of critical wildlife habitat, if sensitive management plans are adopted.

Conservation organizations, realtors and developers have each synthesized their own objectives for natural areas. Combining the property market with a developing market for conservation, they have been able to achieve the development of schemes that afford habitat protection while allowing economic growth. While much is being achieved in converting development into conservation, there remains a distortion in the conservation market in the form of real estate tax, which acts as a disincentive to the protection of natural areas.

Personal taxes

Chapter 4 explained how individuals contributing to conservation can enjoy deductions and exemptions from certain personal taxes. Such tax benefits provide substantial financial rewards and thereby incentives for conservation. However, there are certain disadvantages to using taxes as media for incentives: in particular, inequities associated with providing fiscal incentives through progressively-rated personal taxes. For example, under present federal and most state income tax laws, an individual who donates land or an interest in land to a conservation organization can deduct the value of the gift from his or her taxable income. In such cases, the *true* value of the deduction to the individual depends upon his/her annual gross income. As a result, donors with larger gross incomes reap significantly higher benefits from this practice than those with modest incomes.

For example Assume that individual 'Y' has a gross annual income of $130 000 and pays 36% federal income tax. Chapter 4 provides an example where individual Y donates a conservation easement over some agricultural land to a local land trust. The potential deduction for income tax purposes is $200 000. Y is able to deduct up to $39 000 (30% of his gross income) each year for six years from his federal taxable income liability, until the value of the easement donated ($200 000) has been fully deducted. At 36%, this represents a tax saving of $72 000. Using the same scenario, suppose that individual 'Z' donates a similar area of land to a conservation organization. The potential income tax deduction is also valued at $200 000. Z's gross annual income is only $30 000 and his top rate of tax is 28%. Therefore, Z can deduct only $10 000 (30% of his gross income) each year for the next six years from his federal taxable income liability. At a 28% income tax rate this represents a tax saving of only $16 800.

Differential tax savings in donating a conservation easement are not limited to income taxes. Without the capital gains tax incentives offered for the donation of conservation easements, some landowners would find it difficult to pass land on to their children. This is particularly true if there has been a large increase in the capital value of the land since acquisition, either because the land has been held for a long time or because the land is situated on an urban fringe and has become valuable development land. Essentially, the greater the rise in the capital value of the land since acquisition, the greater the saving to the landowner in donating a conservation easement.

An incentive structure based on tax deductions raises obvious equity issues, but may also affect the *efficiency* with which private land is gathered by conservation organizations for protection. Gifts of rights in land can be expected from landowners with high gross incomes but the land gifted may not always be of high conservation quality. On the other hand, land of high conservation quality may be owned by people with lower gross incomes, who have less incentive to offer land for protection. Conservation organizations are at risk of accepting too many conservation easements from one particular subset of landowners: the high income earners.

Bill Long of the Montana Land Reliance refuses to accept that conservation easements are only 'tax dodges' for wealthy landowners: 'The motive factor in every one of our 38 easements we hold has been for philanthropic reasons – in landowners seeing values they care for and they want to protect ... we call the tax benefit the icing on the cake.' (*Bozeman Daily Chronicle*, 1991). Nevertheless, surveys of conservation easement donors tend to support the observation that tax incentives provide a major motivation for donation. In its 1985 survey, the Land Trust Alliance tried to draw some conclusions about the type of person willing to donate easements. 'The most common easement donor is an individual (or couple) over the age of 50, having a relatively high income which derives from sources unrelated to the land being placed under easement.' (Land Trust Alliance, 1985). Only 4% of reported dealings were with individuals who work the land with the easement and rely on it for income. Potential income tax deduction was the primary motivation for 22% of conservation easement donors surveyed. It was ranked as the second most important motivation (after protecting the land) by 54% of those surveyed. Other motivations were rated much lower by participants but included (in order of importance): estate or gift tax savings, neighbors also granting easements, property tax reductions and prevention of conflict among heirs.

The above analysis suggests that it is vital that conservation organizations are proactive in seeking particular properties and habitats for conservation management and protection: by reacting to approaches made by private landowners, they may accept land of a lower conservation value. Without knowledge of each individual's tax status, conservation organizations cannot know the full cost of accepting a gift of land or property rights: while it may incur costs itself, the costs of providing incentives for the landowner are borne by the nation in terms of lost tax revenue. This situation provides limited information on the *opportunity cost* of protecting one area of land over another. Unfortunately, since the cost of providing the tax incentive for the landowner is borne by other taxpayers (in the form of higher taxes to compensate for the lost revenue), the conservation organization has little incentive to improve its knowledge about the opportunity costs of protection. Most conservation organizations, wishing to enhance their public profile and to seize on opportunities to protect land of some conservation value, will readily accept donations of property rights from these landowners.

If a conservation organization has strict criteria upon which it can evaluate different pieces of land, and so ensure that it only accepts land which is consistent with its own objectives, then decision making is undoubtedly improved. The Nature Conservancy has recognized the costs of holding land and easements in terms of stewardship responsibilities and has stated that it must 'channel resources toward the most biologically significant projects by minimizing spending on sites with less biological significance or sites that can be protected by others.' Some land trusts, particularly those with specific goals or a targeted geographic area, such as the Appalachian Trails Lands, are also likely to ensure that they do not accept land that is not directly consistent with their conservation objectives. However, other conservation organizations, particularly those still at an early stage of establishment and eager to become recognized, may accept land with questionable conservation status.

While the inequalities present in incentives provided through deductions and exemptions from personal taxes may be almost over effective in encouraging landowners to offer their land for conservation management and protection, the nature of land tax in the USA may positively discourage conservation.

Land taxes

United States real estate tax laws are assessed and collected at state level and are determined by the value of the property. In most states, a property with potential for subdivision and development will attract a higher assessment value and so result in higher taxes. This is not conducive to encouraging landowners to preserve wildlife habitat. Calculating the tax with reference to the *developed* value of the land encourages landowners to *realize* that value. In the 1970s, Virginia initiated a taxation based on fair market value. Since most of the coast of Virginia is attractive to development, the value of farmland along the coast increased dramatically: 'Virginia law makers could hardly have designed a better lever to wrest coastal land out of the hands of farmers and into the hands of developers.' (Badger, 1990).

Some sort of land tax relief is needed to redress the balance of this perverse incentive structure. 'Use value farmland assessment' is the assessment of farmland based on its current use (for any of a variety of 'agricultural' purposes) rather than on its market value (Rademacher, 1989). The particular operations for which a farm use can be assessed include a variety of agricultural, livestock, horticultural and aquacultural practices, as defined by each state statute. As a policy, use-value assessment recognizes that the capital value of farmland has risen in the USA at a greater rate than farm incomes. Despite farm real estate being ordinarily assessed at a lower rate of tax than nonfarm real estate, farm property taxes in recent years have been absorbing 7% of farm incomes (double the comparable figure for urban property owners). This preferential tax assessment should change the incentives available to landowners contemplating development. By its very nature, the reduced assessment is most likely to benefit landowners on the urban fringe.

Not all states have adopted use-value farmland assessment. Each state establishes its own rules and regulations. Often, a minimum number of acres are required to be devoted to the particular agricultural operation, or a minimum dollar amount or percentage of income derived from the operation. In addition, taxpayers may have to affirm use for a minimum number of years. Enforcement of the regulations can be quite stringent. Some state governments establish soil fertility grades, factors or values to assess whether the farm is being managed productively or is just a cover for land speculation. Penalties are imposed by 28 states when land is converted from a use-value status to a nonbenefitted use.

The use-value assessment for property tax has two implications for nature conservation. First, it may encourage some landowners, by reducing the property tax burden of the land in its undeveloped state, to preserve areas of land for wildlife habitat. Second, different rules governing whether the assessment will be affected by hunting, fishing or other recreational use of land are used in each particular state. The particular rules will affect the benefits available for land assessed as use-value farmland and subsequently utilized for a fee-paying recreational operation. Erosion of the 'farm-use' status through the introduction of a recreational enterprise could subject the landowner to substantial tax penalties. Penalties can include levying taxes from one to six years before the change of use, levying back taxes (with interest) and imposing a specific land use change tax.

Whether a fee-recreational enterprise will affect the property tax status of the land depends on each state's relevant statutes. In Oregon, the relevant statute (Oregon Revised Statutes) confirms that a farm use will not be affected by hunting, fishing, camping or other recreational use of the land provided that the recreational use of the land does not interfere with the use of the farmland. However, ORS 308.390 states that if there is a change in the farming activity for any reason (which could include activities of a fee-recreation enterprise resulting in a partial reduction of land directly involved in the farming practice or a complete halt of the farming practice), the farm-use assessment will be disqualified on that portion of land directly involved in the nonfarming activity (McClelland *et al.*, 1989).

One way in which a landowner might reduce property taxes on land used for wildlife recreation purposes would be to grant a conservation easement over the land to some qualified conservation organization, forgoing development rights over the land. Land subject to an easement will not usually be reclassified so as to afford a decreased valuation solely because of the creation of the easement. However, the presence of the restrictions on the land should reduce the capital value of the land and so reduce property taxes payable.

Clearly, this is not an option for all landowners: some may not be willing to work with a conservation organization, others may not have land of high enough conservation value to attract a conservation organization's interest. Around half of the USA's states permit 'open space' to qualify as a land use, which attracts use-value assessment. Extension of this privilege to all states might encourage both protection and management of areas for wildlife habitat on private land and increase the benefits available to landowners who may wish to derive income from recreational use of wildlife areas.

Conclusion

Conservation real estate can provide a positive financial incentive for the protection and management of wildlife habitat and natural areas. While partial development may open up the land to human interference, planned management to conservation objectives can help to minimize damage to the environment. Federal and state tax laws may be more effective in encouraging conservation from a particular type of landowner: those with high incomes. Conservation organizations must be wary of this in their selection of properties for conservation easements. Although income tax laws may not encourage donation of conservation easements from lower income earning landowners, current real estate tax laws that place a high capital value on undisturbed property provide a strong incentive for landowners to realize the value assessed through development.

Review of the real estate tax legislation might result in reforms that will remove this institutional barrier to conservation and encourage maintenance of land in its undeveloped state. Alternatively, the government could use an alternative incentive structure to augment conservation behavior. The tax incentives available to private landowners donating property or rights in property to a conservation organization are, effectively, a subsidy to the organization. However, by relying on tax incentives rather than direct cash subsidies, the government cannot know whether it could have used its financial resources to protect other land of a greater area or of a higher conservation value, elsewhere. The use of subsidies to conservation organizations, rather than tax incentives, would enable the government to monitor the exact cost of purchasing protection of natural habitat on private land. In addition, given some evaluation procedure, it would enable both the conservation organizations and the government to assess the costs against the public benefits afforded. From the conservation organization's point of view, better knowledge of the opportunity cost of obtaining protection over a specific area of land may result in more efficient protection. Improved information on the cost of protection might encourage organizations to set priorities and conduct comparative analyzes of the relative benefits of protecting specific areas of land. While in terms of efficiency and knowledge of opportunity costs, subsidies are generally acknowledged as superior to tax incentives, it is debatable as to whether they are more effective as an incentive mechanism. Anyone with an understanding and experience of the farming community will appreciate the degree to which landowners dislike taxes and the lengths to which they will go to avoid payment. The continuation of this inherent aversion to taxes means that tax incentives are likely to provide the most effective means of encouraging conservation behavior in the farming community.

8

Conservation partners

If the private owner were ecologically minded, he would be
proud to be the custodian of a reasonable proportion of such
areas, which add diversity and beauty to his farm and to his
community.

Aldo Leopold (The Land Ethic.)

Conservation partners – imitating the English Estate?

Until now, this book has highlighted the role of conservation organizations
in helping landowners to protect natural areas on private land. In most cases,
the conservation organization involved supplies the legal mechanism for protection, some funding and, most importantly, a commitment to enforcing the
protection of the natural area in perpetuity. In many cases, a conservation
organization will approach a landowner because it perceives that protection
of an area of land is consistent with its objectives. In other cases, a landowner
may approach an organization because he or she would like to protect an
area for private use and nonuse benefits.

In cases where the landowner owns the freehold of all of the land requiring
protection, a successful agreement may be reached. However, Chapter 1 identified that natural areas must be of a sufficient size to protect a wide variety of
wildlife species. In addition, wildlife tend not to recognize private property
boundaries: the policy concerning wildlife on one parcel of land (including
managing habitat and the hunting of wildlife) may affect the amount and quality
of wildlife present on another. There are benefits to be enjoyed where a landowner welcomes the wildlife onto his or her land. However, some landowners
will see the increase in numbers of wildlife as a cost: those areas of agricultural
land close to wildlife habitat are more likely to suffer productive losses as a
result of grazing wildlife. Essentially, the management and land-use practices
of one landowner affect how a neighboring landowner may make use of his or
her land. The degree to which landowners will affect each other may be related
to the size of the parcels of land owned. Clearly, when an area of over say
200 000 acres is held in freehold ownership, adjacent properties may only feel
the effect of the use and management of areas nearer the property's boundary.
As the size of land held decreases, so the possible conflict with neighboring
land management systems might increase.

In England, the effect of such 'neighborly interaction' is often lessened by the presence of the leasehold system. Although relatively small areas of land are farmed as a single holding (the average English farm is around 300 acres), five or six farms might be owned by one individual, together with any surrounding woodland and other nonagricultural natural areas. While each farming unit is let, retention of the freehold of the estate by a single owner enables an overall conservation policy to be enforced in terms of the use and management of land as wildlife habitat.

In the USA, owners of small parcels of private land have, to some extent, imitated the English estate system by grouping together to form more complementary management practices. They have, however, dispensed with the need for the leasehold system and a single owner to bring unity to separate farms. In a manner that is entirely consistent with the American desire for owner-occupation, a group of landowners with similar objectives might gather together to form an organization in their own right in order to improve their immediate environment. By establishing a protection policy that covers several parcels of private land, the private landowners of each parcel effectively create a large reserve without disposing of their individual freehold property rights. In doing so they will seek to protect both the direct and indirect benefits associated with conservation of natural areas by extending protection beyond their own boundaries.

Costs and benefits of co-ordinated management

There are numerous reasons why particular landowners may wish to form a conservation agreement with their neighbors. For example, landowners may not wish to see any further residential development either on their own land or in their immediate environment. In order for each landowner to gain some control over the use of neighboring land, they may form an organization whereby the members are committed to preventing further development on their own land. The need to prevent the destruction of natural areas on neighbouring land may not be only to protect aesthetic and other 'non use' values. For example, overhunting on one area of private land may reduce the quality of hunting on another.

The type of agreement that the landowners will enter into will depend upon the group's requirements. Where protection of natural areas in perpetuity is desired, the landowners may form a formal conservation organization and use conservation easements to protect the land. In other cases, where landowners merely wish to establish a code of conduct for the taking of wildlife or the management of wildlife habitat, a simple agreement may suffice. Research suggests that the more heterogeneous a community, the more difficult coordination becomes (Sugden, 1984). Small groups of landowners with shared ideals, which have identified that they might benefit from collective action, should be able to organize themselves relatively easily.

The benefits that might stem from the collective action of private landowners are numerous. Landowners will enjoy improved recreational use of wildlife habitat, such as hunting, shooting, fishing or merely watching wildlife. In addition, they may enjoy considerable nonuse benefits, such as the benefit of knowing that the area within which they reside and know intimately will be perpetuity in perpetuity. Costs of coordinating management, and of establishing and enforcing codes of practice that embody the landowners' collective objectives for management, need not be expensive. Ostrom (1990), in discussing similar management situations on commons, suggested that shared values between landowners may prevent the need for elaborate monitoring and enforcement mechanisms once the code of conduct has been agreed.

Case study: Leopold Memorial Reserve, Wisconsin

Aldo Leopold is credited as having been one of America's foremost naturalists and conservation writers. He is probably best known as the author of *A Sand County Almanac*, which was published in 1949 and is now an established environmental classic. Born in Burlington, Iowa in 1887, it was not until 1924 that a Forest Service assignment took Leopold to Madison, Wisconsin. In 1933 he became a professor of Wildlife Management at the University of Wisconsin, where he remained for the rest of his life. Around 1935, Leopold purchased an abandoned farm in Sauk County, Wisconsin. The land, which had been logged and farmed in the early twentieth century, was in a poor state. Leopold spent weekends on the farm with his family, staying in a small run-down cabin that they called 'the shack'. While on the farm, the family worked to improve the environmental quality of the land, replanting and restoring particular habitats. The farm provided the setting for the first part of *A Sand County Almanac*, in which Leopold describes beautifully the changing seasons on the farm and discusses examples of man's destructive land use practices (Fig. 8.1). In one of the most significant chapters, Leopold proceeds to a philosophical discussion concerning the environment and makes a plea for the development and adoption of a sound land ethic. The original land purchase is now part of the Leopold Memorial Reserve, a private reserve

Fig. 8.1. 'The Shack'– Leopold Memorial Reserve, Sauk County, Wisconsin. Photograph: Victoria Edwards.

of some 1500 acres in total. The Reserve, which comprises a good deal of cultivated agricultural land, also houses an array of natural but severely damaged habitats, including oak savanna, sedge meadow, prairie types and wetland types. Leopold's first task on purchasing the land was to plant trees merely to restore the vegetation. The farm land had suffered from intensive logging, grazing and plowing. At first the only trees Leopold could afford were white pines and red pines, although within a couple of years planting of trees more characteristic and appropriate to the area was carried out (Haglund, personal communication 1994). Restoration of the land continues under its present management. Apart from the original farm, additional tracts of land have been added to the Reserve but not through the purchase of freehold title to adjacent land. Instead, several neighbouring landowners have agreed to participate in an 'association'.

The agreement

Through a signed agreement, each of the owners participating in the Reserve has agreed to refrain from certain land-use practices that are considered detrimental to the natural quality of the area. While keeping their original freehold property rights, the landowners have agreed to join together to manage their individual parcels of land in a way that is compatible with the objectives of the entire Reserve. The agreement was originally drawn up for a fixed term of five years and to run thereafter until formally terminated by a majority of the landowners party to the agreement. Through the agreement the landowners, in the interest of protecting and enhancing the natural value of the Reserve, have undertaken to:

maintain a co-ordinated trespass control program:

maintain a fire control program;

designate areas not under cultivation as nonagricultural areas, to be kept free from cultivation and grazing;

keep strict control over the granting of any new rights to the land (rights of way, easements, pipelines, etc.) that might open it up to development or threaten the natural areas;

adjust their use of chemicals on the Reserve to protect wildlife habitat, particularly wetland areas;

restrict building and the development of any nonagricultural practices on the Reserve;

agree upon a policy for the cutting of timber on the Reserve and prevent damage to trees; and

employ positive conservation measures to protect and enhance the natural quality of the Reserve.

The agreement is far more proactive in its approach to conservation than many conservation easements. Rather than merely restricting landowner behavior on the Reserve, and so protecting it from outside exploitation, it affords a positive commitment from all parties to actively seek the restoration and enhancement of the environmental quality of the area.

The owners of each parcel of land are represented on the Reserve Management Committee. The activities of the committee are co-ordinated by the Sand County Foundation, a private nonprofit organization which effectively manages the Reserve. In addition, the landowners in the Foundation solicit the advice of a Research Review, a research body made up of scientific experts, to monitor and manage the Reserve as a single unit in their co-ordinated program of research.

More landowners are encouraged to join the association at any time. Rather than push landowners into joining, the Sand County Foundation 'lights a beacon' and hopes that landowners will be attracted to its work: 'The Reserve concept provides land large enough to hold species which a small lot could not sustain, and the innovative and insightful management of the Sand County Foundation is attracting attention of other private landholding entities and individuals who wish to emulate the concept and process' (Coleman & Coleman, 1992).

Research and education

We are engaged in the practice of healing the biologic community, with its human population, on the lands and waters of the northeastern Sauk County, Wisconsin. We do so to provide one model of effective stewardship. This is not to generate precise replicates, but in order to enable other private landowners and committed conservationists to find their own way back to the good earth.

Sand County Foundation, 1989

The Sand County Foundation uses the Reserve as an important vehicle to convey its message of land stewardship. Education on the Reserve extends from primary school visits to invited conservation 'working tours'. At every level, the emphasis is placed on participating in active conservation work on the Reserve. The treatment of visitors is consistent with Aldo Leopold's practical approach to conservation: the Sand County Foundation requires that its visitors contribute to the restoration of the Reserve by working during their visit. Recent projects include the picking and sacking of prairie plant seeds for prairie restoration, the surveying of tracts for weedy plants, such as the European buckthorn, and the subsequent removal of weeds (Fig. 8.2).

The success of the enhancement program to date is quite impressive: the populations of many species of flora and fauna are increasing. In terms of fauna, the Reserve now boasts the presence of bald eagles, otters, wild turkeys and sandhill cranes, amongst other species, all of which were robbed from the area when the land was degraded. In the meantime, continual monitoring of the land, its resources and the conservation practice adopted forms an important long-term experiment in land management. By virtue of direct private action, the landowners contributing to the Reserve have been able to achieve a collective objective of establishing a model on how to enhance ecological quality through good stewardship. It is a testimony to the amount that can be achieved by private landowners through innovative proactive management.

The Leopold Environmental Land Ethic . . . was a sound, voluntary environmentalism that depends on private ownership and stewardship. Much current environmental policy, on the other hand, involves government appropriation of the individual's obligation to improve the land he or she owns. It undermines the very private responsibility that lay at the heart of Aldo Leopold's philosophy and practice.

Coleman & Coleman, 1992

Case study: North Heron Lake, Minnesota

Historically, government intervention in the management of waterfowl in the USA has relied on the acquisition and management of public lands. As early

Fig. 8.2. Conservation work: removing exotic buckthorn, at the Leopold Memorial Reserve. Photograph: Victoria Edwards.

as 1870, Lake Merritt, a state owned refuge near Oakland, California, was established. From the 1920s, about 80 million acres of county, state and federal lands were acquired across the USA to provide waterfowl production, migration and wintering habitats.

Despite a continuing program of public acquisition of land for waterfowl, many populations have declined to the lowest levels ever recorded. Approximately 56% of wetlands in the contiguous states have already been lost and, according to the Fish and Wildlife Service National Wetlands Inventory, the USA will lose an additional 4.25 million acres of wetlands by the year 2000 (US Fish and Wildlife Service, 1990). Declining numbers are due, to a large extent, to the conversion of wetlands to croplands. Given that approximately 87% of overall wetland losses are due to conversions for agricultural purposes, the US Fish and Wildlife Service acknowledges that 'The agricultural community must be a major part of the solution . . .' (US Fish and Wildlife Service, 1990:13).

The importance of directing effort towards encouraging private landowners to conserve wetlands is becoming increasingly apparent. It is estimated that 74% of the USA's remaining wetlands are on private land. A whole host of government programs, federal and state, have been established to provide private landowners with appropriate incentives. In addition, private conservation organizations have initiated programs of their own.

North Heron Lake is situated in southwest Minnesota. The lake provides important breeding, resting and staging habitat for migratory waterfowl. Native Americans named it 'Heron Lake' due to the abundance of black-crowned night herons in the area. The lake is bounded on all sides by private land, some 16 parcels in all, with each landowner having access to the lake. Therefore, the benefits that each landowner may derive from the presence of the lake depend heavily upon the quality of use of the lake by other landowners. In addition, the land use and management practices adopted by each landowner affect the lake's ecology, in terms of the management of the watershed and the run-off of both soil and inorganic substances from the land.

Cooperation between the riparian landowners at North Heron Lake goes back at least as far as the early 1900s. In 1906, landowners surrounding the lake signed 'The Heron Lake Agreement', whereby they agreed to abide by specified shooting hours, restrictions on the use of boats, crossing of open water and 'courtesy'. Essentially, the agreement was designed to avoid disturbing feeding and resting waterfowl and to encourage good shooting practice in an attempt to ensure the conservation of the birds. The original agreement is now on display at the Bell Museum of Natural History in Minnesota and has been cited in numerous books and periodicals.

The lake comprises 3426 acres. While the surface of North Heron Lake itself is public domain, the land beneath the lake and the surrounding lands (which provide direct access to the lake) have been held in private ownership since before the turn of the century. The landowners, therefore, were able to exclude most of the public from the lake and enjoy use of the lake in common. Until recently, the only public access afforded to the lake was via Division Creek. Keen hunters who were determined to gain access to the lake might paddle down Division Creek, since as long as they remained afloat they would not be trespassing on private land. The access via this route is poor, however, and only the most determined hunters would make the effort.

In 1906, the lake was surveyed and boundaries established for each riparian owner from a platform in the open water. The boundaries established were marked by flags and fences and have been respected to the present time (see Fig. 8.3). A primary purpose of the boundaries being established was to define marshland hunting rights and so reduce wildlife disturbance. In 1962 those boundaries were reconfirmed and each of the landowners signed an agreement disclaiming the right to use any of the waters of North Heron Lake that the survey identified as belonging to another owner. Effectively, the riparian landowners divided up a common property resource into private parcels. From 1906, therefore, it has been recognized that no landowner could travel across the lake and each landowner must shoot within his or her own 'slice' of the lake. Commencing in 1962, express written agreement in the form of a license could be obtained to cross boundaries.

Research and wildlife management

As early as 1922 the landowners began to encourage positive management of the area, conducting research into the ecology of the lake and its ability to support waterfowl. There is a hundred-year history of local landowners helping with wildlife surveys of colonial nesting and water birds, as well as ducks and geese.

In 1964, the landowners identified the need to pool their efforts to more efficiently manage the lake as a wildlife habitat and improve the wildlife conditions in all 16 segments. In 1965 the North Heron Lake Game Producers Association was formed and incorporated by charter in 1966. The word 'producers' was chosen because 'it implied improving conditions and carrying out positive action' (Thompson, personal communication 1991). The Association received nonprofit status in 1967. All work carried out for the Association was initially voluntary and unpaid. Various conservation and wildlife pro-

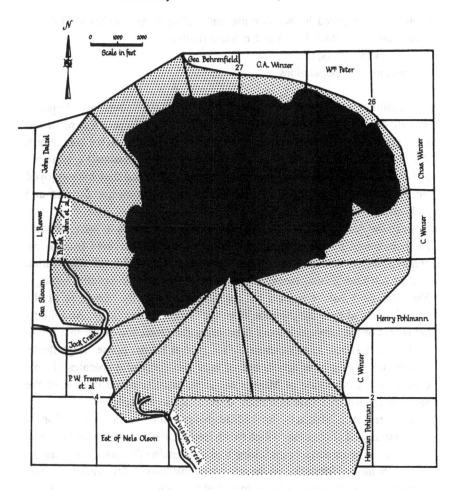

Fig. 8.3. North Heron Lake, Minnesota. From a survey made by C. W. Grove of Windom, Minnesota, July 1909.

duction projects were established, together with several research projects. As regular monitoring of wildlife revealed the status of each species present, research was conducted to establish the cause of any declining populations and appropriate action taken. For example, projects have included:

improving nesting cover on land;
planting trees, native grasses and flowers;
planting and managing roadside prairies;
establishing food plots and feeding wildlife;

allocating cropland for wildlife use and setting aside land for wildlife;
erecting wood duck boxes and mallard platforms;
restoring wetlands; and
managing fish populations.

In addition, self-imposed rules and regulations were adopted to improve wildlife conditions. For example, the shooting of Canada geese was restricted for seven years while geese were released and reestablished at the lake. North Heron Lake Game Producers are now recognized as an outstanding example of North America's efforts to reestablish Canada geese and they have enjoyed similar success with the reestablishment of trumpeter swans, ospreys and gadwalls.

Since the Association was formed, its budget for projects to increase wildlife populations on and around North Heron Lake has increased from a few hundred dollars to over $20 000 per annum. In early 1989, the Association hired a full-time habitat manager with a masters degree in wildlife biology. The manager's salary is paid by contributions from the members of the Association and by a contribution from the Nature Conservancy. An important aspect of the manager's work is ensuring the protection and management of the North Heron Lake sanctuary. This special habitat houses large flocks of colonial nesting birds and is one of only two successful colonies of Forester's Tern in the state of Minnesota. Strict restrictions on the use of the area are needed to ensure minimum disturbance to the colonies.

Status reports produced by the wildlife manager illustrate the variety of species and wealth of populations which North Heron Lake currently supports. They also serve as an important monitoring tool in assessing the quality of management, providing information on the relative populations of different species and assisting in the evaluation of the impact of recreational activity on the Lake. Game species include: ring-necked pheasant (*Phasianus colchicus*), Canada goose (*Branta canadensis*), gray partridge (*Perdix perdix*), eastern cottontail rabbit (*Sylvilagus floridanus*), white-tailed deer (*Odocoileus virginianus*), jackrabbit (*Lepus townsendii*), jack snipe (*Gallinago gallinago*) and mourning dove (*Zenaida macroura*). However, the variety of nongame species protected by the reserve is, perhaps, even more impressive. Nongame species include: Forester's tern, black tern, Franklin's gull, ring-billed gull, herring gull, great blue heron, great egret, black-crowned night heron, western grebe, eared grebe, horned grebe, pied-billed grebe, white pelican, double-crested cormorant, swallow, marsh wren, white-faced ibis, common moorhen, American bittern, red fox, coyote, woodchuck, mink, racoons, opposum, fox squirrel, and badger.

The efforts of the landowners around North Heron Lake have been acknowledged widely as a fine example of private land stewardship. The success of the lake to attract and retain such a wealth of wildlife is due not only to the positive contribution of the management practices of the area, but also to the willingness of the private landowners to respect the wildlife's need for rest and to restrict their own use of the area accordingly.

Conclusion: new threats

It is important to appreciate that not all private conservation initiatives rely on the sale of products to consumers. Some private landowners are willing to protect natural areas because of the benefits they derive from the area in its conserved state. This chapter has explained how the private use and nonuse benefits associated with the ownership and management of natural areas may be sufficient to encourage private landowners to adopt land-use practices that seek to maximize environmental benefits. The chapter has used two case studies to illustrate how a group of landowners might combine, without public sector involvement, to strengthen the protection of their respective pieces of land by imposing some common management policy. The success of such private initiative is heavily dependent upon the the incentives available to the landowners. Any interference within that incentive structure may destroy the willingness or ability of landowners to continue to contribute to wildlife protection. Once the users of the land have devised and established successful institutions to facilitate cooperation in management, external intervention may upset the operation of the management system and so jeopardize its future. In particular, the arrival of a third party who will not adhere to the rules established may, eventually, result in all members of the group reneging on the agreed code of practice. Runge (1984) refers to the 'contingent strategies' of individuals, explaining that observations of how others will behave will have an effect on the strategies adopted by individual members of the group.

Third party intervention

For some time, the Department of Natural Resources (DNR) of the State of Minnesota has expressed a wish to purchase any piece of private land adjacent to North Heron Lake. Such land would allow the DNR to open up the lake to the public. Clearly, the private landowners surrounding the lake have been concerned that this intrusion would destroy their efforts to protect the

wildlife preserve, which has not only suffered relatively little human disturbance but also benefits from extensive wildlife management.

On two separate occasions the DNR were unsuccessful bidders for parcels
of land around North Heron Lake, being outbid by the existing private landowners. In 1991, however, the DNR acquired two parcels of land adjoining
the North Heron Lake System on behalf of the State of Minnesota. While not
abounding the lake, the land affords access to Jack Creek, one of the lake's
main tributaries. Access to the lake via Jack Creek is good, albeit one and a
half miles to the lake. Since acquiring the land, the State's Department of
Natural Resources has indicated its intention to use the land to afford public
access to North Heron Lake and to allow public shooting on the Lake. Certainly, the quality of wildlife on and around the lake would make it a most
attractive proposition for any public hunter considering purchasing a license
from the State. In addition, there is likely to be considerable public pressure
for the DNR to provide public access to the lake as a means of justifying
purchasing the land with public revenue.

Naturally, the Association is concerned that opening up the area to the
public will effectively destroy the benefits of almost a century of careful
habitat management. There can be little doubt that the establishment and
successful enforcement of an agreed code of practice in use of and access to
the lake has provided a strong incentive structure for landowners to manage
the wildlife soundly. There is uncertainty over the effect that public access
to the lake will have in terms of the self-imposed restrictions to use. Each
year the members of the Association agree to a new code of conduct, in
the spirit of the original agreement of 1906. The regulations imposed are
supplementary to federal and state regulations on shooting. They are designed
to ensure not only that shooting practice is tailored to suit North Heron Lake's
specific needs, but also that a better than average standard of wildlife management is achieved.

Ironically, as the private landowners improve the quality of the water at
North Heron Lake, the possibility of the area being accessible to public hunters increases. Minnesota State law prohibits hunters from shooting from
boats unless they are firmly anchored to solid ground. Equally, hunters must
not conceal themselves in any habitat other than permanent or emergent vegetation. Currently, the lack of emergent vegetation around the lake's margins
makes it easier to enforce trespass laws because hunters are forced to step
firmly onto private land. As the water quality of the lake improves and more
light is able to penetrate the margins, emergent vegetation is likely to facilitate hunting without trespass. This exemplifies the dilemma facing landowners who are keen to exercise best management practice and improve the

quality of their habitat for nature conservation yet, by their action, facilitate increased human presence on the land.

Concern over public access is not restricted to the uncertainty of whether the public will conform with the code of conduct at the lake: there is also the issue of contribution to the management of the lake. The amount of effort that the Association's members are willing to put into positive management of the area might be directly linked to: (a) perceptions of future benefits of their efforts and (b) the degree to which individual contribution (whether in the form of labor, materials or financial) is equitable and reflected in subsequent enjoyment of the resource. Sugden (1984) suggested that individuals will feel a moral responsibility to contribute when they know that others are contributing.

Tang (1992) stated that the ability of a group of individuals to design and implement effective institutional arrangements for managing a natural resource is often a reflection of:

1. the amount of information available to resource users;
2. the degree of uncertainty they face;
3. the extent to which opportunistic behavior is employed by the resource users;
4. the frequency with which resource users are expected to make individual contributions to management of the resource.

At North Heron Lake, the Association has produced an annual newsletter for the last thirty years that fulfils two very important tasks. First, the newsletter helps to build a sense of pride in the membership and, second, it reinforces best management practice by making members aware of the success of their concerted action.

In both case studies cited in this chapter, the management practices that have been adopted and implemented have been successful because of: (a) willingness of the resource users to collect and analyze information about the resource and its ability to sustain wildlife over time; (b) the certainty they face over the continued stability of the resource, because the information supplied enables alteration of the code of practice concerning use of the resource according to particularly sensitive periods and places; and (c) the expectation of each individual user that other members of the group will conform with the code of practice and will contribute to positive management of the resource.

The benefits to be enjoyed from the cooperation of private landowners at North Heron Lake extend beyond each individual landowner's lifetime. Many of the present owners are descendants of the first-time resident farmers who drew up the original agreement. It is likely that the present owners will bequeath the title to their land, and the benefits of North Heron Lake which accompany it, to their own children. The continuity provided by private land ownership no doubt encourages the strong conservation ethic which exists at North Heron Lake.

The North American Waterfowl Management Plan was signed by United States and Canadian officials in 1986 as a commitment to wildlife conservation. In the USA the Plan encourages cooperation between federal and state governments and conservation agencies. Joint ventures between public and private organizations are also sought. The 1990 Wetlands Action Plan states that

It is not economically feasible to protect wetlands from agricultural impacts through wetlands acquisition alone. However, agricultural losses can be reduced significantly by encouraging agricultural practices that conserve wetlands and by seeking new opportunities to protect wetlands through easements, leases, economic incentives, regulatory improvements, and the voluntary participation of private individual.

US Fish and Wildlife Service, 1990

It seems ironic, therefore, that State intervention at North Heron Lake (where arguably *no* further incentives for conservation are required), might well upset the successful land-use practices and water management already exercised. It is possible that in the case of North Heron Lake, the most valuable contribution that the State of Minnesota can make to the management of waterfowl is to abide by the practice of management initiated by the landowners. It has, after all, been recognized as successful for almost a century. Recent meetings between the landowners and the DNR suggest that the Department acknowledges the success of North Heron Lake as a wildlife area and attributes much of that success to the private landowner initiatives. Ironically, however, it would seem that having purchased land which could physically provide public access to the lake, the DNR may have difficulty in refusing to open it up to the public. The land was purchased with public revenue and despite the DNR's recent pledge to maintain protection of wildlife at the lake they may be forced to compromise protection of the area. Now that the land is in the State's ownership it will be difficult, politically, to keep the area closed to the public. In respect of the wildlife at North Heron Lake, an immediate sale of the land in question to a private landowner may be the most effective and efficient means of protecting this special area.

9

Towards a more holistic approach

In general, the trend of the evidence indicates that in land, just
as in the human body, the symptoms may lie in one organ and
the cause in another. The practices we now call conservation
are, to a large extent, local alleviations of biotic pain. They are
necessary, but they must not be confused with cures. The art of
land doctoring is being practiced with vigor, but the science of
land health is yet to be born.

Aldo Leopold (Wilderness)

The market for nature conservation

Chapter 1 of this book explains the importance of the conservation of wildlife
and habitat on private land. The sheer size of the private domain dictates
that it is essential we understand and improve the means of protecting natural
areas on such land. Traditionally, the benefits associated with the protection
of natural areas have been regarded as 'nonrival' (one person's enjoyment
has no effect on the enjoyment of another) and 'nonexcludable' (it is impossible to exclude people from enjoying the benefits). Assumptions of this type
lead to a bias towards government supply of nature conservation and/or
government regulations over private land. Chapter 2 stresses that natural areas
are capable of producing a whole range of benefits, some of which can be
captured and a charge levied. It explains that by exercising control over
access to the benefits to be enjoyed from natural areas, private landowners
can produce goods and services that are compatible with the protection of a
natural area and also allow them to recover some of the costs of conservation.

Frequently, the nature conservation market requires agreements that will
secure protection of a natural area, while allowing some type of good or
service to be developed in association with the area. Chapter 4 explains that
a variety of formal and informal protection mechanisms facilitate such protection and product development, while chapters 5 and 6 illustrate how such
agreements can be encouraged and how financial benefits, derived from a
fee-paying enterprise on private land, may be sufficient to ensure that the
land is managed for improved wildlife habitat for both game and nongame
species. Income incentives of this type will be of greater importance when the
opportunity cost of conservation increases. Provided with a choice between
intensive agriculture or nature conservation, fee-paying recreational
enterprises may actually provide sufficient financial benefits to persuade a
landowner to protect the land in its natural state. However, the determination

of these benefits will need to take into account both the increased costs of developing and operating it and the foregone benefits of the alternative land use. If development of the enterprise can be achieved with low capital input and little increase to the fixed costs of the holding, then income from the fee-paying enterprise will contribute to a more substantial increase in the holding's overall budget.

In many cases the option to protect land and manage it for nature conservation purposes will be evaluated against alternative uses that could possibly provide greater financial incentive. Chapter 7 explains how the economic benefits of natural areas land can be capitalized when the land is sold as a conservation property with full or partial development restrictions. Partial development, in particular, offers landowners the prospect of a financially viable compromise when the opportunity cost of managing the land in its natural state may be too high for them to contemplate.

The actors in the market place

Conservation organizations, working with private landowners and the public, act as valuable agents in transactions. They have created:

1. a conservation market place in which transactions can take place;
2. a range of products through which it is possible to recover the costs of conservation;
3. a range of tools that facilitate protection of natural areas while allowing for their economic use.

In doing so, they have created a market for nature conservation that is constantly being refined to facilitate the protection of more sites.

Landowners also have an important role to play in the market for nature conservation. An agenda for the future provision of conservation areas will continue to evolve under the guidance of those policy makers representing the 'public' interest. Such an agenda is likely to follow the traditional approach to conservation in which governmental and quasi-governmental agencies are charged with the responsibility of representing the public objectives for conservation. However, by approaching conservation in a planned, proactive and *private* manner, there is the potential for landowners to seize public values for their own private benefit while enhancing the natural qualities of their land. First, landowners must identify areas which may benefit from being managed in a way that will conserve and enhance their special natural features. These natural conserved areas must then be marketed to the general public. Landowners in America have a tradition of providing for consumer demand: for two centuries they have satisfied America's food market. More recently they have played a vital role in developing new products, from processed food substances to recreational enterprises. Through a history of meeting and indeed sometimes stimulating public demand for products, landowners are well placed to adapt a similar approach to satisfying public demand for conservation. Whereas government agencies will target incentives to areas which *they* perceive to be of special interest or under particular threat, landowners can deal directly with the public to provide conservation.

Advantages of the market

The market for nature conservation has five distinctive qualities that help it to contribute to the enhancement of natural areas.

1. **Information** The market is an effective means of providing information for example on habitat characteristics, threats to habitats and the values people place on conservation. Such information is important not only in relation to the benefits of protecting particular sites but also to the costs of doing so.
2. **Economic Awareness** The market stimulates awareness of the opportunity costs of protecting particular areas. Buyers in the conservation market are continuously forced to make decisions over resource allocation and where to spend their next conservation dollar.
3. **Trading** The market provides a place for the trading of property rights, enabling conservation buyers to improve their aggregate protection.
4. **Competition** The market encourages competition between players. Competition among conservation organizations should result in the reduction of the costs of accomplishing the protection of an area. When resources are limited and must be competed for, organizations are aware that for each site protected there may be other areas towards which funds cannot be directed. The knowledge of this opportunity cost attached to each transaction encourages organizations to keep the cost of conducting transactions to a minimum, allowing them to achieve the largest protected land base possible. Conservation agencies, therefore, have an incentive to increase their own economic efficiency: increased efficiency allows them to satisfy their members demand for conservation by protecting more sites (Anderson, 1982). They can market their achievements as two-fold: the amount of benefit they have acquired by protecting a stated acreage of land *and* the efficiency with which they have achieved their goal. The Nature Conservancy does just this in its annual report, informing members

of the cumulative area protected and the proportion of funds spent directly on conservation acquisitions as opposed to administrative costs (The Nature Conservancy, 1993).

5. **Flexibility** The conservation market is sufficiently flexible to allow players to respond to new opportunities. Individual landowners must react quickly to changing demands, since natural areas often do not become threatened until impending transactions such as sale or development arise. While conservation organizations may not perceive themselves as entering into a real estate market, in reality they must often work within that market to achieve their goals. Ideally, the institutional arrangement of the conservation market needs to give conservation organizations (a) the freedom to conduct trading in the property rights to natural areas and (b) the ability to act on an equal footing with other competitors for land. Given the dynamics of such a market, conservation organizations need to use a variety of tools for exchanging freehold and other property rights.

Improving the market

If the conservation market is to be used as a policy tool, we must seek ways of improving it. In particular, when imperfections are perceived analysts should question *why* the market may not be functioning properly. Part of the uniqueness of the private land conservation movement worldwide is that it operates apart from government. It must be recognized, however, that governments do influence the incentives and choices available to individuals through their interaction in conservation institutions. There are always characteristics of any market situation that act as disincentives to transactions. Such disincentives must be identified and neutralized for efficient operation of the market.

If a market is to rely on a range of products from a diverse range of producers it must have a variety of legal options at its disposal for conducting transactions. It is crucial, therefore, that legal mechanisms can be developed by which desired transactions can take place. The government plays an important role in developing such transaction mechanisms. Anderson & Leal (1991) have placed certain prerequisites on the attainment of market solutions to environmental problems: 'only when rights are welldefined, enforced and transferable will self-interested individuals confront the tradeoffs inherent in a world of scarcity. As entrepreneurs move to fill profit niches, prices will reflect the values we place on resources and the environment.'

As a general rule, the more property rights that can be separated from the 'bundle', the more transactions can be tailored to meet individual requirements. In particular, the ability to purchase only those property rights that are essential for natural area protection and management of the land allows conservation organizations to use their funds efficiently; the purchase of the freehold interest in land is often an expensive luxury. Easements created by statute, allowing perpetual protection, are widely available for use by conservation organizations in the United States. Easements enable organizations to prevent the development of a natural area through the negotiation of a restrictive covenant over the land limiting future land use. A similar tool is employed in New Zealand and is referred to as a 'statutory covenant' (Edwards & Sharp, 1990). In Britain, however, the only private organization to hold such statutory power is the National Trust; all other private conservation organizations have no such powers at their disposal and are unable to divide up property rights in this way.

Expanding the role of the private sector

Historically, the private conservation market in the USA has emphasized the protection and management of relatively small areas of biological diversity. By finding ways of funding the ownership and management of such areas, individuals and groups have enabled a wealth of wildlife areas to remain intact. Essentially, the private sector has been building up an ark of natural habitat and species. More recently, however, organizations have recognized the need to adopt a more holistic approach to nature conservation. Rather than isolate natural areas and protect them as small islands of diversity, a holistic approach seeks to integrate the management of natural areas with economic use of neighboring land. It stems from a maturing attitude to conservation that has been forced to acknowledge the importance of sustainable economic use of lands.

This approach is typified by the Nature Conservancy's Virginia Coast Reserve. The Virginia Eastern Shore has been designated a World Biosphere Reserve, one of 12 in North America. It is a United Nations designation, reserved for the world's most important ecosystems, defined as large multipurpose areas intended to protect functioning ecosystems, while finding ways to incorporate human activity which will not degrade the area. While recognizing the term 'Biosphere Reserve' as one means of referring to large scale ecosystems, the Conservancy uses its own term of the 'Last Great Places' and in 1991 launched an initiative 'to work with local communities to protect entire ecosystems, not just isolated islands of biodiversity' (Sawhill, 1991). Significant biological resources are protected in a core natural area. In an adjoining buffer zone, local people develop a productive and sustainable economic base through the ecologically sound use of their natural resources.

One of the reasons the Eastern Shore was selected as a Biosphere Reserve was the long term example it serves of humans living alongside natural communities. Extensive research will be carried out to learn about the natural systems and to monitor any environmental changes that may occur. In particular, records will seek to find land use practices, such as farming and fisheries, that will improve the 'well-being' of the community without degrading the natural environment. Agriculture and fisheries have always dominated the economy of the Eastern Shore. Once home to the Algonquin Indians, a group of colonists established a fishing and saltmaking camp on the islands in the early seventeenth century. The mainland was later settled by farmers and livestock raisers. In the late nineteenth century several exclusive hunting clubs were established on the islands. The area is currently the leading vegetable, soybean and grain producing region in Virginia and an important region for tourism and sport fishing. In addition, a multimillion dollar seafood industry has been established on the Eastern Shore.

This more holistic approach to land management is not peculiar to the Nature Conservancy. Smaller conservation organizations are also attempting to broaden the focus of their work. For example, in 1972 the Appalachian Trail Conference resolved 'to seek and establish an Appalachian Greenway encompassing the Appalachian Trail and a significant width to provide a nationally significant zone of dispersed types of recreation, wildlife habitat, scientific study, and timber and watershed management, as well as to provide vicarious benefits to the American people.' (Trust for Appalachian Trail Lands, 1990). The proposal was adopted in 1990 as 'A Greenway Initiative', seeking to add a 'countryside zone' to the corridor of the Appalachian Trail. The greenway initiative takes the work of the Trust to new ambitious heights and the objectives of the Trust (see Trust for Appalachian Trail Lands, 1991) in pursuing the Greenway Initiative include:

1. identifying resources and specific tracts of land in need of protection;
2. building work coalitions with other individuals, organizations and agencies working to protect open space in the Appalachians;
3. taking inventory of federal, state and local government plans and programs that can be used to preserve open space in the Appalachian Mountains;
4. monitoring and participating in local land-use planning and zoning activities that could enhance protection of the greenway;

5. promoting statewide and local initiatives that enhance open space and greenway protection;
6. educating the open space advocacy community and government agencies on the goals of the greenway initiative; and
7. expanding its own land acquisition programs.

The objectives signify a move away from a policy emphasis of direct acquisition, to a more advocacy role for the Trust. It is very similar to the change in character of the Nature Conservancy's work, from the protection of specific targeted areas of land as preserves, to a broader focus on regional environmental protection.

It seems logical that a more holistic approach to environmental protection, looking at habitat rather than species preservation and regional rather than local land protection, demands the cooperation of private organizations with public agencies and local communities. The North Heron Lake Game Producers Association has identified the need to work with other organizations. The Association has recognized that the quality of its own 4000 acre habitat depends heavily on not only the management of the entire watershed, but also the quality of water coming through that watershed from adjacent lands. In addition, the visiting wildlife depend upon even more remote lands for their existence. In this respect, the Association has begun to liaise with national and international organizations outside its locale, such as Ducks Unlimited, the Nature Conservancy, the Aldo Leopold Institute, Pheasants Forever and the International Game Foundation.

Whereas in its preserve protection the Conservancy was recognized as independently pursuing conservation objectives through direct land protection, it appears that in its attempt to gain a wider perspective on ecosystem protection it has established the need to work with many more groups. Historically an independent organization, the Conservancy has begun to recognize that 'to achieve effective conservation we need to work with a coalition composed of three entities: the government, the private sector and other conservation organizations. Any hope of protecting our natural lands and waters is futile without public sector participation.' (The Nature Conservancy, 1990a). Partnerships are sought with local landowners, individuals, civic groups, the chamber of commerce, businesses, foundations, public agencies and other nonprofit organizations to create a local community that will work to fulfill mutual economic, social and environmental goals. In essence, the Conservancy is seeking to implement a planning system. The Conservancy's policy objective for management of the Virginia Eastern Shore is stated as 'the development of protective planning to promote balanced zoning and land use laws.' In 1991 Northampton County (one of the counties included in the Reserve) passed a comprehensive land-use plan, supporting environmental protection while allowing appropriate economic growth. Further such plans are sought from other counties.

In both of the above cases, progression from a policy of 'building an ark' to a more holistic approach to land use and management seems to have coincided with the attainment of the conservation organization's primary goals. In both the case of the Nature Conservancy and that of the Trust for Appalachian Trail Lands, the primary purpose of the organization had, to a large extent, been satisfactorily achieved. While few people would suggest that the Conservancy had completed its task of preserving America's biological diversity, there can be little doubt that it has made a substantial contribution to the protection of natural areas on private land. It seems natural, therefore, that the Conservancy should set it sights on further horizons and it is doing so by extending (a) the geographic area of its targeted protection and (b) its own brief in terms of influencing the management of *all* lands.

The costs of expansion

Like the Nature Conservancy, the Trust for Appalachian Trail Lands has been very successful in securing land, in its case for the Appalachian Trail. When an organization has a track record of success in its environmental program, it seems vital that the organization builds on that success and fixes its sights on new horizons. However, expansion may come at a price. It is possible that in seeking a broader and more holistic approach to conservation, private sector organizations are in danger of destroying the very essence of their success – their independence from government.

Recently, the Nature Conservancy has expressed concern at the overwhelming task of extending the protection of species and natural communities throughout the USA. Such concern may stem from the recent broadening of its horizons: the Conservancy has committed itself to protecting natural areas in Latin America and the Pacific region. In addition, the Conservancy has become acutely aware of the need to properly manage land once it has been obtained for its portfolio. In its conservation strategy (The Nature Conservancy, 1990), the Conservancy identified the need for each state to review its entire preserve portfolio and set a timetable for defining and funding adequate stewardship of all preserves retained by the Conservancy:

Growth in our stewardship program has not kept pace with the increase in our preserve portfolio. Several states with significant preserves have no stewardship staffing. Resources in many others are absorbed by the needs for basic property management.

Stewards often spend the majority of their time on facilities maintenance, property rights and preserve visitors leaving little or no time to work directly to protect the biological elements of the preserve.

In addition, it recognized the need to coordinate stewardship functions in a way that will influence the stewardship of biologically important areas on other public or private land. Understandably, therefore, in its 'Conservation Strategy for the Nineties', the Conservancy identified the need to 'channel resources towards the most biologically significant projects by minimizing spending on sites with less biological significance or sites that can be protected by others.'

Private, going public?

Partnership agreements for managing public lands are an obvious way for the Conservancy to extend protection of America's natural areas. However, further developments have suggested that the Conservancy's ability to portray itself as a purely private organization, respecting private property rights, may be at jeopardy.

As a private nonprofit organization, the Conservancy derives most of its revenue from its 600 000 membership fees, individual contributions and grants from corporations and charitable foundations. In 1991 contributions of this type totalled $51.7 million, with other revenue coming from investment income ($14.7 million) and 'other income' in the form of sales of goods, etc. ($23.4 million). However, in the same year the Conservancy obtained an additional $83.1 million of revenue from the sale of natural areas to state and federal agencies (an increase of $40.3 million on the previous year) (The Nature Conservancy, 1991a).

The practice of selling land on to government agencies is a significant source of income for the Conservancy and provides it with the means to reinvest in property. However, it is a practice that has been criticized for several reasons. First, there is a danger that the Conservancy's membership and financial supporters will soon perceive the Conservancy as an agent for the government. Since its 40 years of success has depended very much on its private sector approach to conservation, this could weaken support for the organization from its members. Second, there is widespread recognition that most landowners prefer not to deal with government agencies when discussing or negotiating land-use and management practices. The Conservancy has relied heavily on the willingness of landowners to contribute property or rights in property for the protection of biological diversity. This has involved landowners gifting land to the Conservancy, gifting easements, or entering into informal 'registry' agreements. Suspicion of the Conservancy's involvement with government agencies, coupled with recognition that any land gifted to the Conservancy may be transferred to the public domain, may reduce the willingness of landowners and the public to support and work with the Conservancy. Indeed, one Wisconsin family refused to enter into the informal registry program because it believed that in having its land on a Nature Conservancy register made it a target for compulsory purchase by a government agency. Other landowners in Wisconsin have refused to sell land to the Conservancy when they have admitted that the land will be resold to the state (Braker, personal communication 1991).

Suspicions may not be based on pure prejudices. The Conservancy recognizes the need for sound stewardship of land in order to protect properly its natural resources and yet in its 'Conservation Strategy for the Nineties', the Conservancy admits that 'work on our preserves has been used to inspire the appropriate management of public lands. However, we have been unable to ensure proper stewardship of all projects transferred to government agencies.' Essentially, the change in the institutional arrangement of the Conservancy's operation may result in a change in the incentive structure of the landowners and members it depends upon. The derivation of a consumer's enjoyment of a conservation product is most complex and unique to each consumer. However, it seems likely that many of the benefits that consumers derive from conservation products are associated with the fact that they are *privately* produced. The willingness to contribute to private conservation funds may be based on perceptions of belonging to an exclusive group (Olson, 1965) and recognition that public funds are insufficient to supply the amount of land that they would like to see protected (Weisbrod, 1975). Private organizations should be cautious, therefore, of crossing the public/private boundary. Once consumers perceive an element of public intervention they may no longer wish to contribute to the private organization's funds or participate in its activities, as the element of exclusivity is lost and public involvement suggests that public funds should be collected fairly from all.

A role for government?

While this chapter warns that private organizations must be aware of the implications of public sector partnerships, it does not advocate a public/private dichotomy. In fact the Conservancy has been successfully involved with the public sector for over 15 years through its inventory program, and recognition of the need to work with the public sector in its own stewardship program has resulted in a number of initiatives on public land. In 1989 the

Conservancy signed an agreement with the US Department of Defense (which controls 25 million acres of land) to allow the Conservancy to undertake biological inventories and management plans on 900 of its defense installations. In 1990 it signed a similar agreement with the Bureau of Land Management to establish a partnership to protect rarities on the Bureau's 200 million acres (one eighth of the land mass of continental USA). Most recently, a 'memorandum of understanding' has been drawn up between the Conservancy and the US Forest Service, allowing 'cooperative conservation' in 18 national forests in California. It is hoped that the project will be extended to a national conservation agreement with the Forest Service.

The success of such partnerships suggests that organizations like the Nature Conservancy have built up a bank of skills and expertise in natural areas management that should be protected, nurtured and exploited. Along with that expertise, conservation organizations have built up a tremendous reserve of goodwill amongst landowners and their members. For these reasons, the government might seek to make even better use of the willingness of private landowners to contribute to nature conservation. Each chapter of this book attempts to identify areas where there is a role for government to augment the work of the private sector conservation market by improving the institutional arrangement of that market. In addition, the government could recognize the efficient operation of private sector conservation organizations and support their work through the allocation of public funds. The government can reduce the problem of the free-rider and reduce some of the organizational costs of revenue collection through the imposition of taxes. Such funding can be important in enabling private organizations to expand their geographic terms of reference and seek a more holistic approach to land management. At North Heron Lake, public as well as private partnerships are being established: 'If we ignore opportunities for liaising with the public sector we will be washed over by the millions invested in the North American Waterfowl Management Plan' (Tony Thompson, personal communication 1994).

However, the public *funding* of conservation does not necessarily call for public *provision*. Competitive tendering by private conservation organizations for public funds would be a welcome introduction to the nature conservation market. It is a practice that has been introduced in New Zealand with some notable success. Encouraging voluntary protection is a cost-effective way of providing for the protection of wildlife habitat on private land. It builds on the basic willingness of private landowners to provide for wildlife, so ensuring that the quantity and quality of protected habitat approaches socially optimal levels. It encourages the personal involvement of landowners in conservation and increases their awareness of the natural qualities of their land. One fundamental constraint in retaining areas in their natural state is that the environmental benefits associated with conservation tend to be long term, while shorter term financial gain can be achieved through alternative land-use practices. In the decision over whether to exploit the land for alternative uses, private landowners have a greater incentive than any public body to take into account the needs and desires of future generations. When discounting the future benefits to calculate the present value of conserving an area in its natural state, governments will respond to the current voters and place a higher value on current potential benefits. In contrast, because the land is likely to stay in the hands of their family, to be enjoyed by their children and grandchildren, private landowners will place a higher value on the future benefits commonly associated with conservation.

In addition, the *voluntary* protection of habitat places less pressure on effective enforcement mechanisms and so reduces the present costs of conservation. Since the landowner has chosen to protect the wildlife habitat voluntarily, one might assume that he or she has a greater interest in ensuring correct management for its protection, rather than relying on enforcement by formal procedures. The effect of introducing involuntary conservation has been demonstrated by the problems of operating the Endangered Species Act 1973. The Act has enjoyed limited success in restoring many species to a sustainable level. Of the hundreds of species added to the list (the list topped 1000 species in 1990), only five have recovered sufficiently to be removed from the list (Griffith, 1993). Problems associated with the Act have reportedly included the problems of over-anxious property owners preempting its enforcement by removing or destroying listed species and those likely to be listed, rather than face the restrictions imposed by the Act (Griffith, 1993). Similar problems were encountered in Britain with the restrictions imposed by the designation of an area as a 'Site of Special Scientific Interest' (SSSI) under the Wildlife and Countryside Act 1981. The Act, until a subsequent amendment plugged the loophole, effectively afforded property owners a three month period in which to remove legally any special species from their land before restrictions were formally imposed.

Conclusion

There are over 5000 protected areas of some sort in the world today, including almost 800 internationally protected sites, covering 4% of the world's land area (World Resources Institute, 1990). However, while the extent of designated protection may appear impressive, the effectiveness of that protection is questionable. Whelan (1991: 10) commented that 'the protection is often only on paper, due to both a lack of funds and local support. . . . often countries (*sic: governments*) focus their attention on purchasing land, but fail to follow up with adequate funds for infrastructure and management.'

For some time now, the world has recognized that it must save all the 'cogs and wheels' of the earth's wealth of resources. However, it has only recently recognized, as Leopold states, that the 'science of land health' must be adopted and action needs to be taken in a holistic manner to protect whole ecosystems rather than merely species. This does not mean that special sites will be protected and individual species ignored. Definitions of the term 'biodiversity' tend to refer to multiple levels of biological organization (genetic diversity, species diversity and ecosystem diversity) and emphasize the variety and variability contained within each level (OTA, 1987).

The objective of protecting biodiversity demands, by its very nature, a holistic and integrated approach to resource management. There is a temptation to suppose that only national and international public bodies hold the power to achieve such an ambitious task. Such an assumption ignores three essential features of land management in the USA:

1. 60% of the land area is in private ownership;
2. private landowners must derive some sort of economic return from their land; and
3. protection, like all land management, is a dynamic not static process and involves protecting the diversity of a site from natural and unnatural destruction.

Recognition of these three features forces us to acknowledge that no public agency can protect biological diversity in isolation: it would be politically, economically and physically impossible.

Private approaches to conservation play a vital role in the protection of biodiversity and to abandon them would be environmental suicide. Encouraging private efforts of conservation reinforces a superior land ethic by rewarding the wise use of natural resources. The development of a sound land ethic is probably the most important step in any country's biodiversity agenda.

References

Allen, W. H. (1988). Biocultural restoration of a tropical forest. *Bioscience*, **38**: 156–61.

Anderson, T. L. (1982). New resource economics: old ideas and new applications. *American Journal of Agricultural Economics*, **64**: 759–92.

Anderson, T. L. & Leal, D. R. (1991). *Free Market Environmentalism*. Pacific Research Institute for Public Policy, San Francisco, CA.

Appalachian Trail Conference (1990). The Appalachian Trail in New England: Supplemental Lands Protection Project. ATC, Harpers Ferry, WV (unpublished).

Baden, J. & Stroup, R. (1983). *Natural Resources: Bureaucratic Myths and Environmental Management*. Ballinger Publishing Co., Cambridge, MA.

Badger, C. J. (1990). Eastern Shore gold. *Nature Conservancy* **40**(4): 7–15. The Nature Conservancy, Arlington, VA.

Bandow, D. (1986). A new approach for protecting the environment. In *Protecting the Environment: A Free Market Strategy*. The Heritage Foundation, Washington, DC.

Barnett, S. (1985). *New Zealand in the Wild*. Collins, Auckland.

Belt, D. C. & Vaughn, G. F. (1988). *Managing Your Farm for Lease Hunting and a Guide to Developing a Hunting Lease*. Extension Bulletin No. 147, Delaware Cooperative Extension, College of Agricultural Sciences, University of Delaware, Newark.

Berglas, E. (1976). On the theory of clubs. *American Economic Review*, **66**: 116–21.

Bozeman Daily Chronicle (1991). State Ranked Fourth in the Nation in Land Easements, p. 21, 13 October, 1991.

British Birds (1982). Editorial: Codes for rarity finders and twitchers. *British Birds*, **75**: 301–3.

Brock, J. *et al.* (1990). *Montana Tourism Marketing Research Project*. Montana State University, Bozeman.

Bromley, P. (1989). Wildlife opportunities: species having management and income potential for landowners in the East. In *Natural Resource Management and Income Opportunity Series: Alternative Enterprises*, ed. A. Ferrise & W. N. Grafton, R. D. No. 754, pp. 1–8. West Virginia University Extension Service, Morgantown.

Brunner, P. (1982). Conservation Real Estate. In *Private Options: Tools and*

168

Concepts for Land Conservation, ed. B. Rusmore, A. Swaney & A. D. Spader, pp. 73–8. Island Press, Covelo, CA.

Bryan, W. (1991). Ecotourism on Family Farms and Ranches in the American West. In *Nature Tourism: Managing for the Environment*, ed. T. Whelan, pp. 75–85. Island Press, Covelo, CA.

Buchanan, J. M. (1965). An economic theory of clubs. *Economica* **32** (125): 1–14.

Buchanan, J. M. (1973). The Coase Theorem and the Theory of the State. *Natural Resources Journal*, **1973**: 579–94.

Burton, J. (1978). The externalities, property rights and public policy: private property rights or the spoilation of nature. In the Epilogue to *The Myth of Social Cost* by S. N. S. Cheung. Institute of Economic Affairs, London.

Chase, A. (1987). *Playing God in Yellowstone: The Destruction of America's First National Park*. Harcourt Brace Jovanovich, San Diego, CA.

Chesapeake Bay Foundation & University of Maryland School of Law, Law Clinic (1988). *A Guide of Private Land Protection in Maryland*. Chesapeake Bay Foundation, Annapolis, MD.

Cicchetti, C. J. & Freeman, M. (1971). Option Demand and Consumer Surplus: Further Comment. *Quarterly Journal of Economics*, **85**: 528–39.

Coase, R. H. (1960). The Problem of Social Cost. *Journal of Law and Economics*, III: 1–44.

Coleman, R. & Coleman, T. (1992). Growing with the Leopold land ethic. *Philanthropy* VI(II): 4–6.

Connolly, G. E. (1980). Predators and predator control. In *Big Game of North America, Ecology and Management*, ed. J. L. Schmidt & D. L. Gilbert, pp. 369–94. Stackpole Books, Harrisburg, PA.

Conrad, J. M. (1980). Quasi-option value and the expected value of information. *Quarterly Journal of Economics*, **94**: 813–20.

Council on Environmental Quality (1972). National parks. In *Environmental Quality*. The Third Annual Report of the Council on Environmental Quality. US Government Printing Office, Washington, DC.

Council on Environmental Quality (1984). *Environmental Quality 1987–1988*. The Fifteenth Annual Report of the Council on Environmental Quality. US Government Printing Office, Washington, DC.

Council on Environmental Quality (1985). *Environmental Quality 1987–1988*. The Sixteenth Annual Report of the Council on Environmental Quality. US Government Printing Office, Washington, DC.

Council on Environmental Quality (1986). *Environmental Quality 1987–1988*. The Seventeenth Annual Report of the Council on Environmental Quality. US Government Printing Office, Washington, DC.

Council on Environmental Quality (1988). *Environmental Quality 1987–1988*. The Eighteenth and Nineteenth Annual Report of the Council on Environmental Quality. US Government Printing Office, Washington, DC.

Craighead, F. C., Jr (1979). *Track of the Grizzly*. Sierra Club Books San Francisco, CA.

Cutler, M. R. (1989). Appreciative use of wildlife – the recreational choice of three out of four Americans. In *Natural Resources Management and Income Opportunity Series: Philosophy and Policy of Recreational Access*, ed. W. N. Grafton & A. Ferrise, R. D. No. 762, pp. 1–8. West Virginia University Extension Service, Morgantown.

Deacon, R. & Johnson, M. B. (1985). *Forestlands: Public and Private*. Ballinger Publishing Co., Cambridge, MA.

Diehl, J. & Barrett, T. S. (ed.) (1988). *The Conservation Easement Handbook:*

Managing Land Conservation and Historic Preservation Easement Programs. Land Trust Alliance, Washington, DC & The Trust for Public Land, San Francisco, CA.

Dowdle, B. (1981). An institutional dinosaur with an axe: or, how to piddle away public timber wealth and foul the environment in the process. In *Bureaucracy vs Environment: The Environmental Costs of Bureaucratic Governance*, ed. J. Baden & R. Stroup, pp. 170–85. The University of Michigan Press, Ann Arbor.

Dunlap, T. R. (1988). *Saving America's Wildlife: Ecology and the American Mind, 1850–1990*. Princeton University Press, Princeton, NJ.

Edington, J. M. & Edington, M. A. (1986). *Ecology, Recreation and Tourism.* Cambridge University Press, Cambridge.

Edwards, V. M. & Sharp, B. M. H. (1990). Institutional arrangements for conservation on private land in New Zealand. *Journal of Environmental Management* **3**: 313–26.

Eggertsson, T. (1990). *Economic Behaviour and Institutions.* Cambridge University Press, Cambridge.

Faro, J. B. & Eide, S. H. (1974). Management of McNeil River State Game Sanctuary for nonconsumptive use of Alaskan brown bears. *Proceedings of the Annual Conference of the Western Association of State Game and Fish Commissioners*, **54**: 113–18.

Foresta, R. A. (1984). *America's National Parks and Their Keepers.* Resources for the Future, Washington, DC.

Fox, S. (1981). *The American Conservation Movement: John Muir and His Legacy.* The University of Wisconsin Press, Madison.

Gallatin Valley Land Trust (1994). Here come the conservation easements! *Gallatin Valley Land Trust Newsletter*, **4**(1): 1.

Glick, D. (1991). Tourism in Greater Yellowstone: Maximising the Good, Minimising the Bad, Eliminating the Ugly. In, *Nature Tourism: Managing for the Environment*, ed. T. Whelan, pp. 58–74. Island Press, Covelo, CA.

Goldsmith, F. B. (1974). Ecological effects of visitors in the countryside. In *Conservation in Practice*, ed. A. Warren & F. B. Goldsmith, pp. 217–31. John Wiley & Sons, London.

Gregory, R., Mendelsohn, R. & Moore, T. (1989). Measuring the benefits of endangered species preservation: from research to policy. *Journal of Environmental Management*, **29**: 399–407.

Griffith, J. J. & Knoeber, C. R. (1986). Why do corporations contribute to The Nature Conservancy? *Public Choice*, **49**: 69–77.

Griffith, V. (1993). Halting Industry in its Tracks. *Financial Times* (London), p. 9, 10 February 1991.

Halliday, T. (1978). *Vanishing Birds, their History and Conservation.* Sidgwick & Jackson, London.

Hildebrandt, R. (1989). *Public Access to Private Kansas Lands for Recreation.* Report of Progress 582, Agricultural Experiment Station, Kansas State University, Manhattan.

Hilts, S. & Moull, T. (1986). *Natural Heritage Stewardship Program: 1986 Annual Report.* University of Guelph, Ontario.

Israel, S. & Sinclair, T. (ed.) (1987). *Indian Wildlife.* Insight Guides, APA Productions (HK) Ltd, Singapore.

Krutilla, J. V. & Fisher, A. C. (1975). *The Economics of Natural Environments: Studies in the Valuation of Commodity and Amenities Resources.* Resources for the Future/Johns Hopkins University Press, Baltimore, MD.

Kwong, J. (1987). Public and private benefits: the case for fee hunting. *PERC Viewpoints* No.2. Political Economy Research Center, Bozeman, MT.

Land Trust Alliance (1985). Report on the 1985 national survey of government and nonprofit easement programs. *Land Trust's Exchange* 4 (3) (December). Land Trust Alliance, Washington, DC.

Land Trust Alliance (1989). *Statement of Land Trust Standards and Practices.* Land Trust Alliance, Washington, DC.

Land Trust Alliance (1990). *Land Trusts: Finding the Answers that Save Land.* Publicity brochure, Land Trust Alliance, Washington, DC.

Land Trust Alliance & National Trust for Historic Preservation (1990). *Appraising Easements: Guidelines for Valuation of Historic Preservation and Land Conservation Easements*, 2nd edn. Land Trust Alliance, Washington, DC.

Land Trust Exchange (1985). *The Federal Tax Law of Conservation Easements.* Land Trust Alliance (formerly Land Trust Exchange), Washington, DC.

Leopold, A. (1933). *Game Management.* Charles Scribner's Sons, New York.

Leopold, A. (1949). *A Sand County Almanac.* Oxford University Press, New York.

Leopold, A. (1953). *Round River.* Oxford University Press, New York.

Lindberg, K. (1990). *Tourism as a Conservation Tool.* World Resources Institute, Washington, DC.

Louisiana Private Lands Team (1989). *Louisiana's Private Lands Initiative: Supporting the North American Waterfowl Management Plan*, US Fish and Wildlife Service, Washington, DC.

Maryland Environmental Trust (1989). *Conservation Easements: To Preserve a Heritage.* Annapolis, MD.

McClelland, S. D., Cleaves, D. A., Bedell, T. E. & Mukatis, W. A. (1989). *Managing a Fee-Recreation Enterprise on Private Lands.* Extension Circular 1277/March 1989, Oregon State University Extension Service, Corvallis.

McLean, I. (1987). *Public Choice: An Introduction.* Basil Blackwell, Oxford.

McNeely, J. & Miller, K. (eds.) (1984). *National Parks Conservation and Development: The Role of Protected Areas in Sustaining Society*, Proceedings of the World Congress on National Parks. Smithsonian Institution Press, Washington, DC.

McNeely, J. (1988). *Economics and Biological Diversity: Developing and Using Economic Incentives to Conserve Biological Resource.* International Union for the Conservation of Nature and Natural Resources, Gland, Switzerland.

Michener, J. A. (1985). *Texas.* Random House Inc., New York.

Moment, G. B. (1970). Man-grizzly problems – past and present: implications for endangered species. *Bioscience* **20**: 1142–4.

Montana Land Reliance & Land Trust Exchange (1982). *Private Options: Tools and Concepts for Land Conservation.* Island Press, Covelo, CA.

Myers, N. (1986). Tropical Deforestation and a Mega-Extinction Spasm. In *Conservation Biology – The Science of Scarcity and Diversity*, ed. M. Soulé. Sinauer Associates, MA.

Myers, N. (1988). Tropical forests: much more than stocks of wood. *Journal of Tropical Ecology* **4**: 209–21.

Myers, N. (1989). *Deforestation Rates in Tropical Forests and their Climatic Implications.* Friends of the Earth, London.

National Shooting Sports Foundation. (1987). *Hunting 'Frequency and Attitude' Survey Summary.* NSSF, Riverside, CT.

Newman, J. R. & Schereiber, R. K. (1984). Animals as indicators of ecosystem responses to air emissions. *Environmental Management*, **8**: 309–24.

Newmark, W. D. (1987). A land-bridged island perspective on mammalian extinctions in western North American parks. *Nature*, **325**: 430–2.

Newport News (1986). Partners in Protection, Editorial, May 18, 1986.

Olson, M. (1965). *The Logic of Collective Action: Public Goods and the Theory of Groups*. Harvard University Press, Cambridge, MA.

Ophuls, W. (1973). Leviathan or oblivion. In *Toward a Steady State Economy*, ed. H. E. Daly, pp. 215–30. Freeman, San Francisco, CA.

Organisation for Economic Cooperation and Development (1989). *Renewable Natural Resources: Economic Incentives for Improved Management*. OECD, Paris.

Ostrom, E. (1986). An agenda for the study of institutions. *Public Choice*, **48**: 3–25.

Ostrom, E. (1990). *Governing The Commons*. Cambridge University Press, Cambridge.

OTA (US Congress, Office of Technology Assessment) (1987). *Technologies to Maintain Biological Diversity*. US Government Printing Office, Washington, DC.

Owen, C. N., Wigley, T. B. & Adams, D. C. (1985). Public use of large private forests in Arkansas. *Transactions of the North American Wildlife and Natural Resources Conference* **50**: 232–41.

Owen, C. N. (1989) Recreational use on industrial forest ownership: problems and opportunities. In *Natural Resource Management and Income Opportunity Series: Legal Issues*, ed. A. Ferrise & W. N. Grafton, R.D. No. 744, pp. 30–36. West Virginia University Extension Service, Morgantown.

Paterson, J. H. (1989). *North America: A Geography of the United States and Canada*, 8th edn. Oxford University Press, Oxford.

Pearce, D. W. & Turner, R. K. (1990). *Economics of Natural Resources and the Environment*. Harvester Wheatsheaf, London.

Prescott-Allen, C. & Prescott-Allen, R. (1986). *The First Resource: Wild Species in the North American Economy*. Yale University Press, New Haven, CT.

President's Commission on the American Outdoors (1987). *The Legacy, Challenge*. Island Press, Washington DC.

Rademacher, J. J. (1989) Protective legal measures and concerns of private landowners. In *Natural Resource Management and Income Opportunity Series: Legal Issues*, ed. A. Ferrise & W. N. Grafton, R. D. No 774, pp. 36–51. West Virginia University Extension Service, Morgantown.

Raven, P. (1988). Our diminishing tropical forests. In *Biodiversity*, ed. E. O. Wilson. National Academy Press, Washington, DC.

Rhoads, S. E. (1985). *The Economist's View of the World: Government, Markets and Public Policy*. Cambridge University Press, Cambridge.

Runge, C. F. (1984). Institutions and the free rider: the assurance problem in collective action. *Journal of Politics*, **46**: 154–81.

Sand County Foundation (1989). Leopold Memorial Reserve: restoring processes in the landscape. *Newsletter*, 1989. The Sand County Foundation, Madison, WI.

Satchell, J. E. (1976). *The Effects of Recreation on the Ecology of Natural Landscapes*. European Committee for the Conservation of Nature and Natural Resources, Council of Europe. Natural Environment Series 11.

Sawhill, J. C. (1991). To dream and to care. *Nature Conservancy* **41**(3): 3 The Nature Conservancy, Arlington, VA.

Schoenfeld, C. A. & Hendee, J. C. (1978). *Wildlife Management in Wilderness*. Wildlife Management Institute & Boxwood Press, Pacific Grove, CA.

Schultze, C. L. (1977). *The Public Use of Private Interest*. Brookings Institution, Washington, DC.

Scott, D. (1988). The American Story. In *For the Conservation of the Earth*, pp. 202–6. ed. V. Martun, pp. 202–6. Fulcrum Inc., Golden, CO.

Shaw, W. W. & Mangun, W. R. (1984). *Nonconsumptive Use of Wildlife in the United States: An Analysis of Data from the 1980 National Survey of Fishing, Hunting and Wildlife-Associated Recreation.* US Department of the Interior Fish and Wildlife Service, Resource Publication 154 Government Printing Office, Washington, DC.

Simonds, G. (1988). Incentives affecting wildlife recreation in the public land states. In *Recreation on Rangelands, Promises, Problems, Projections,* Symposia Proceedings of the Society for Range Management, ed. D. Rollins, pp. 74–6. Corpus Christi.

Singer, P. (1975). *Animal Liberation, a New Ethics for Our Treatment of Animals.* Avon Books, New York.

Small, S. J. (1988). *Preserving Family Lands: A Landowner's Introduction to Tax Issues and Other Considerations.* Power & Hall Professional Corporation, Boston, MA.

Smith, A. (1776). *An Inquiry into the Nature and Causes of the Wealth of Nations.* Oxford University Press, Oxford.

Soutiere, E. C. (1989). Waterfowl: income and potential problems. In *Natural Resource Management and Income Opportunity Series: Fish and Wildlife Management* A. Ferrise & W. N. Grafton, R.D. No. 752, pp. 1–11. West Virginia University Extension Service, Morgantown.

Steinbach, D. W. & Ramsey, C. W. (1988). The Texas lease system: history and future. In *Recreation on Rangelands, Promises, Problems, Projections.* Symposia Proceedings of the Society for Range Management, ed. D. Rollins, pp. 54–68. Corpus Christi, TX.

Sugden, R. (1984). Reciprocity: the supply of public goods through voluntary contributions. *Economic Journal,* **94**: 772–87.

Tang, Shui Yan (1992). *Institutions and Collective Action: Self-Governance in Irrigation.* Institute for Contemporary Studies Press, San Francisco, CA.

Thatcher, P. (1988). A world wilderness inventory. In *Conservation For the Earth,* ed. V. Martin, pp. 357–61. Fulcrum Inc., Golden, CO.

The Nature Conservancy (1955). *Operating Manual.* The Nature Conservancy, Arlington, VA.

The Nature Conservancy (1989) *Annual Report.* The Nature Conservancy, Arlington, VA.

The Nature Conservancy (1989a). *Voluntary Private Land Registration.* Publicity brochure, The Nature Conservancy, Wisconsin Chapter, Madison, WI.

The Nature Conservancy (1990). *A Conservation Strategy for the 1990's: the Nature Conservancy's Strategic Plan.* The Nature Conservancy, Arlington, VA.

The Nature Conservancy (1990a). *1990 Annual Report.* The Nature Conservancy, Arlington, VA.

The Nature Conservancy (1991). *The Islands: the Virginia Coast Reserve.* The Virginia Coast Reserve, The Nature Conservancy, Brownsville, Nassawadox, VA.

The Nature Conservancy (1991a). *1991 Annual Report.* The Nature Conservancy, Arlington, VA.

The Nature Conservancy (1992). Detectives of diversity, *Nature Conservancy,* **42**, (1): 22–7, The Nature Conservancy, Arlington, VA.

The Nature Conservancy (1993). *1993 Annual Report.* The Nature Conservancy, Arlington, VA.

The Nature Conservancy (undated). *Gifts of Land,* Publicity Publication, The Nature Conservancy, Arlington, VA.

The Trust for Appalachian Trail Lands (1990). Greenway concept to guide future trust activities. In *Trail Lands: The Newsletter of the Trust for Appalachian Trail Lands,* **7**(1): 1.

The Trust for Appalachian Trail Lands (1991). The Appalachian greenway challenge, by Shaffer Dennis. In *Trail Lands: The Newsletter of the Trust for Appalachian Trail Lands* **7**(2): 23.

Tjaden, R. L. (1989). An urbanising perspective: use of forest and woodland alternatives. In *Natural Resource Management and Income Opportunity Series: Alternative Enterprises* ed. A. Ferrise & W. N. Grafton, R.D. No. 749, pp. 1–15. West Virginia University Extension Service, Morgantown.

Trefethen, J. B. (1975). *An American Crusade for Wildlife.* Winchester Press, New York.

US Department of Agriculture (1986). *Agricultural Statistics, 1986* National Agricultural Statistics Service, Washington, DC.

US Department of the Interior, Fish and Wildlife Service & US Department of Commerce, Bureau of the Census (1993). *1991 National Survey of Fishing, Hunting and Wildlife-Associated Recreation.* Government Printing Office, Washington, DC.

US Fish and Wildlife Service (1990). *Wetlands Action Plan.* US Department of the Interior, Washington, DC.

Wall Street Journal (1990). The Market Conservation's Best Friend, by Stroup, Richard L., 19 April 1990.

Walsh, R. G., Bjonback, R. D., Aiken, R. A. & Rosenthal, D. H. (1990). Estimating the public benefits of protecting forest quality. *Journal of Environmental Management,* **30**: 175–89.

Weeden, R. (1976). Nonconsumptive users: a myth. *Alaska Conservation Review.* **17**(3): 3–15.

Weisbrod, B. A. (1964). Collective consumption services of individual consumption goods. *Quarterly Journal of Economics,* **78**: 471–7.

Weisbrod, B. A. (1975). Toward a theory of the voluntary nonprofit sector economy. In *Altruism, Morality and Economic Theory* ed. E. S. Phelps, pp. 171–95. Russell Sage Foundation, New York.

Whelan, T. (ed.) (1991). *Nature Tourism: Managing for the Environment.* Island Press Covelo, CA.

White, P. S. & Bratton, S. P. (1980). After preservation: philosophical and practical problems of change. *Biological Conservation,* **18**: 241–55.

Wilkes, B. (1977). The myth of the non-consumptive user. *Canada Field-Nat,* **91**: 343–9.

Willard, B. E. & Marr, J. W. (1970). Effects of human activities on alpine tundra ecosystems in Rocky Mountain National Park, Colorado. *Biological Conservation,* **2**: 257–65.

Willard, B. E. & Marr, J. W. (1971). Recovery of alpine tundra under protection after damage by human activities in the Rocky Mountains of Colorado. *Biological Conservation,* **3**: 181–90.

Willis, K. G. (1989). Option value and non-user benefits of wildlife conservation. *Journal of Rural Studies,* **5**: 245–56.

Willis, K. G., Benson, J. F. & Saunders, C. M. (1988). The impact of agricultural policy on the costs of nature conservation, *Land Economics,* **4**: 147–57.

Wilson, E. O. (1988). The current state of biodiversity. In *Biodiversity,* ed. E. O. Wilson. National Academy Press, Washington, DC.

World Commission on Environment and Development (1987). *Our Common Future.* Oxford University Press, Oxford.

World Resources Institute (1990). *World Resources 1990–91*. Oxford University Press, Oxford.

Wright, B. A. (1989). Towards a better understanding of recreational access to the nation's private lands: supply, determinants, limiting factors. In *Natural Resource Management and Income Opportunity Series: Supply of Recreation*, ed. A. Ferrise & W. N. Grafton, R. D. No. 758, pp. 1–14. West Virginia University Extension Service, Morgantown.

Index